# Studies in Systems, Decision and Control

Volume 401

The series "Studies in Systems, Decision and Control" (SSDC) covers both new developments and advances, as well as the state of the art, in the various areas of broadly perceived systems, decision making and control–quickly, up to date and with a high quality. The intent is to cover the theory, applications, and perspectives on the state of the art and future developments relevant to systems, decision making, control, complex processes and related areas, as embedded in the fields of engineering, computer science, physics, economics, social and life sciences, as well as the paradigms and methodologies behind them. The series contains monographs, textbooks, lecture notes and edited volumes in systems, decision making and control spanning the areas of Cyber-Physical Systems, Autonomous Systems, Sensor Networks, Control Systems, Energy Systems, Automotive Systems, Biological Systems, Vehicular Networking and Connected Vehicles, Aerospace Systems, Automation, Manufacturing, Smart Grids, Nonlinear Systems, Power Systems, Robotics, Social Systems, Economic Systems and other. Of particular value to both the contributors and the readership are the short publication timeframe and the world-wide distribution and exposure which enable both a wide and rapid dissemination of research output.

Indexed by SCOPUS, DBLP, WTI Frankfurt eG, zbMATH, SCImago.

All books published in the series are submitted for consideration in Web of Science.

More information about this series at http://www.springer.com/series/13304

Wojciech Rafajłowicz

# Learning Decision Sequences For Repetitive Processes—Selected Algorithms

Springer

Wojciech Rafajłowicz
Faculty of Information and Communication
Technology
Wrocław University of Science
and Technology
Wrocław, Poland

ISSN 2198-4182 ISSN 2198-4190 (electronic)
Studies in Systems, Decision and Control
ISBN 978-3-030-88395-9 ISBN 978-3-030-88396-6 (eBook)
https://doi.org/10.1007/978-3-030-88396-6



This Springer imprint is published by the registered company Springer Nature Switzerland AG
The registered company address is: Gewerbestrasse 11, 6330 Cham, Switzerland

# Contents

# List of Figures

# List of Tables

# Chapter 1
# Introduction

Methods and algorithms for learning decision sequences in a repetitive process can be helpful in achieving optimal results under certain constraints.

In this book, it is planned to present the theory and algorithms for a large class of processes that include not only production processes, but also other areas such as computer systems and large simulation studies. One current example is the spread of COVID-19 that serves in this book as a testing example for several algorithms.

Decision-making is an interdisciplinary study area, where attempts are made on theoretical studies of the process and methods are proposed for achieving the best possible solutions for problems and processes.

Repetitive processes are a class of problems currently under research. In such a process, its nature provides us with multiple subsequent passes that have the same mathematical model. One good example is additive manufacturing, where multiple passes are made in laser cladding. Another is the production of identical parts. When decisions can be made from pass-to-pass, one can use this opportunity to improve the process behavior or its final outcome.

Learning is one of the features of intelligence, so it is also considered as the most important part of artificial intelligence (AI). The methods that can learn help in solving difficult problems with high computational complexity, including

- the traveling salesman problem, which has to be solved many times to endlessly learn varying road conditions, arising, e.g., due to road rebuilding,
- the repeating planning problem,
- board games,
- classifying image sequences,

among others. Possible applications also include optimization by extensive simulations of large-scale processes such that probing a goal function is highly expensive. An example of such processes is the growth of cancer cells.



W. Rafajłowicz, *Learning Decision Sequences For Repetitive Processes—Selected Algorithms*, Studies in Systems, Decision and Control 401,
https://doi.org/10.1007/978-3-030-88396-6_1

Most of the existing approaches to computer-aided decision making concentrate on individual decisions. Moreover, many approaches to optimizing decision sequences have been developed over the course of the last eighty years. In particular, the optimal control theory of discrete-time systems provides a large number of methods and algorithms such as those based on the celebrated Bellman's dynamic programming and its further extensions. Observe, however, that the bulk of approaches provided by control theory concentrates on improving decisions along a current pass (run) of the process, taking into account a present state and present and future decisions only. An exception in control theory is the stream of research known as iterative learning control, well known under the acronym ILC (see Sect. 7.2 for a brief review). This approach was one of the inspirations for writing this book. A closer look at the literature revealed that an approach known as run-to-run control (Sect. 7.1) is based on the idea of improving decision sequences between passes (runs). At present, similarities and differences between ILC and run-to-run control are known, but they are still studied.

More classic approaches, namely, stochastic approximation that is experiencing a renaissance (Sects. 6.1, 6.3) and the response surface methodology (Sect. 6.4) can be interpreted as algorithms, which are based on learning from pass to pass.

Finally, algorithms developed by researchers working on the optimization algorithms (local and global) achieved the level of efficiency that allows us to solve problems with hundreds of variables, using a relatively small number of goal function eveluations. Hence, they can also be considered as tools for optimizing decision sequences, when interpreting improving iterations as learning between passes of the process. A common look at all these approaches to learning decision sequences for repetitive processes is one of the objectives of this book.

In Table 1.1, a summary of a selection of learning algorithms, depending on a priori knowledge or assumptions, concerning the problem at hand, is provided. The leading column is the length of a decision sequence to be taught. The second column specifies the type of the optimization problem and our knowledge about a model of a process. Namely, the model free case means that only observations of a goal function are available, the model-based approach assumes that at least a rough model is available (e.g., as a quadratic function), which is made more accurate from pass to pass. Finally, the model supported approach means that a model is available, but it is only partly used for computing an improving direction of search, while remaining information is collected from observations of a real process. Abbreviations and acronyms in the third column have the following meaning:

**diff. evol.+filter** —the differential evolution algorithm equipped with a population filter for handling constraints,
**SPADL** —Simultaneous perturbation algorithm for decision learning,
**RSM** —algorithm based on response surface methodology,
**grad. along pass** —algorithm based on computing the gradient of an objective function simultaneously for all decisions along a pass.

The analysis of Table 1.1 suggests that problems which are difficult due to the need of searching a global optimum with constraints of a multi-modal function available

**Table 1.1** Suggestions for selecting a learning algorithm, depending on the length of a decision sequence and assumptions (explanations in the text)

| Length | Assumpt. | Algorithm | Sec. |
|---|---|---|---|
| Short | Multi-modal, model-free | diff. evol.+filter | 4.2 |
| Medium | Local, model-free | SPADL | 6.2 |
| Medium | Local, model-based | RSM | 6.3 |
| Long | Convex, model-supported | grad. along pass | 7.4 |

through its observations only, can be solved for relatively short decision sequences. If it suffices to locate a local minimum only, then longer sequences can be taught less accurately by the SPADL algorithm or more exactly by the RSM approach, but at the expense of confining to shorter sequences of a medium length. As one may expect, long sequences can be taught when the problem is convex and we have both a model and observations.

From the viewpoint of a priori information about the process to be optimized one can—roughly—distinguish the following classes of problems:

**model-free** approaches to optimization, characterized by the requirement that only observations of values of the goal function are available as the process response to sequences of decisions,

**model-based** optimization, which fully relies on a model,

**model-supported** or model-inspired optimization that is partly based on a model, which is, however, considered as either a rough approximation of the process behavior or as an insufficiently accurate one.

The well-known class of model-free problems is searching for the minimum of a regression function when only its random realizations are available, e.g., values of the goal function plus random errors, as is typical for the class of methods known as the stochastic approximation. It is worth mentioning that a process manifesting itself only through goal function values at selected points (decision sequences) can have complicated dynamics or spatiotemporal dynamics, nonlinearities etc.

Model-based approaches to optimization are the most spread and already highly developed. They include a wide range of problems, starting from seemingly simple, unstructured problems of searching for the optimum of a goal function that is available either as a formula or a simple computational procedure, and ending on highly structured problems, which require solving sets of ordinary differential equations or partial differential equations to obtain one value of a goal function for one sequence of decisions. An additional challenge arises when we are looking for a sequence of decisions that provides the global minimum of a multimodal global function. In this book we concentrate mainly on this aspect, putting emphasis on handling constraints, especially when they are not explicitly available as simple formulas.

The possible role of model-supported (or inspired) approaches to learning decision sequences seem to be underestimated. The well-known class of algorithms that fall into this stream of research is the response surface methodology developed by the research community working on production quality monitoring. The leading idea is to use simplified (linear or quadratic) models to learn more accurately decisions or process parameters providing better quality of products. As proposed in Sects. 7.3 and 7.4 this idea can be generalized for learning long decision sequences under constraints in the form of difference equations. The essence of this generalization is in replacing a model response by observations of a real process, but using the model for estimating a descent direction, which is the steepest descent when the model is correct.

In addition to a priori information, the role of stimulating the process by a decision sequence and recording its reactions is emphasized in this book as the main source of learning information. The next ingredient is the methodology of learning that is mainly based on estimating a descent or the steepest descent directions and performing steps along them. Notice, however, that gradients (or their estimates) are calculated simultaneously for all the decisions along each pass. In other words, improvements are made synchronously for all the decisions along a pass. This is in contrast to local improvements of current decisions only. This approach allows us to write down almost all the algorithms considered in this book in a unified way. This includes also the differential evolution method that can also approximately be viewed as gradient type algorithm.

From the practical point of view, the problems of decision sequence learning for repetitive processes can be divided into three types.

**Short sequences** that can be taught with simple tools like traditional optimization methods.

**Medium-sized sequences** require more advanced tools like stochastic gradient descent or differential evolution.

**Long sequences** where problems can be solved using iterative methods that are based on a model of a process, which allows for evaluating a descent direction of the goal function.

Images form a specific group where decisions are made based on images or a series of images. The image as a data source requires specialized methods due to stream intensity (hundreds of megabytes per second), which makes on-line decision making a challenge.

The COVID-19 pandemic can be seen in the context of learning as an iterative process. The subsequent waves can (unfortunately) be seen as a repetitive process, so we can adjust the mitigation strategy between them. Obviously public sentiment and the economy impose constraints (Chap. 5). At the beginning of writing this book, the author had the hope that when this book is published the COVID-19 pandemics mitigation scenarios remain only scholarly illustrative examples, but the reality has occurred to be much less optimistic.

The book has the following structure. In Chap. 2 a general overview of the notation is presented along with some basic ideas.

In Chap. 3 basic concepts and algorithms regarding learning are provided.

In Chap. 4 a brief overview of differential evolution and the concept of the filter is introduced as a method of handling constraints.

Subsequently, in Chap. 5 the basic model of the spread of COVID-19 is presented. The mitigation problem is considered as a decison problem. Symbolic calculations are considered. The differential evolution with filter is used to solve this problem.

A stochastic gradient is described as a learning method in Chap. 6. The method of the iterative learning decision sequence is presented in Chap. 7.

The special case of learning from image sequences is presented in Chap. 8.

At present, theories and algorithms of decision support and optimization systems have so many approaches and rich literature that it is completely impossible to discuss them even briefly in one book. Therefore, the selection of the methods and algorithms presented here is confined mainly to those for which the author has certain contributions and/or experience. Additionally, necessary background material and bibliographical remarks are presented.

By necessity, many interesting problems are left outside the scope of this book. They in particular, include

- multicriteria (multiobjective) optimization (see [40, 172, 185]), the exception being Chap. 4, in which the idea of handling constraints is considered from the point of view of multiobjective optimization,
- fuzzy sets approach to decision making [26, 65] as well as neuro-fuzzy approaches [150],
- soft computing techniques [151],
- pattern recognition and machine learning—nonparametric approach [36] and syntactic reasoning [107] (an exception is Chap. 8),
- theory of games [90].

In addition, many computational tools and algorithms were left ouside the scope of this book, in particular

- dynamic programming and reinforcement learning (see [10, 171, 182]),
- expert systems,
- intelligent computing [86] and symbolic computations (see Sect. 5.3 for an application related to COVID-19 mitigation),
- artificial neural networks and their learning although many stochastic gradient algorithms (Chap. 6) can directly be applied for these purposes,
- engineering diagnostics [73],
- fuzzy sets [71]

An inspiration for writing this book comes from iterative learning control (ILC). The author has contributed to this area of research [138, 140]. A closer look at the ILC revealed that in many research areas one can also find learning algorithms for repetitive processes, although this aspect is frequently not mentioned explicitly. Thus, the leading idea of this book is to collect most of the existing approaches to learning decision sequences, with the exception of reinforcement learning, and

- to present them in a standarized way, allowing for comparisons,
- assessing them from the point of view of their applicability for shorter and longer sequences.

Other contributions of this book can be summarized as follows:

- the differential evolution algorithm is extended by adding a population filter that allows for incorporating complicated constraints in the process of learning,
- testing this algorithm on a simple, but still unexpectedly accurate model of COVID-19 and comparing it with the Newton method implemented in an non-classic way, namely, by using a hybrid of symbolic and numerical methods of calculating directions of improvements,
- an algorithm of learning longer decision sequences is proposed for processes with a memory,
- the domain of classic learning from vectors of observations is extended to images and the algorithm for learning a recognizer to classify image sequences is proposed and tested on real-life data of a laser-based, additive manufacturing process.

The emphasis in this book is put on learning algorithms. Their convergence properties are also briefly discussed. This book will be of interest to researchers, PhD and graduate students in computer science and engineering, operations research and decision making. It can be also of certain interest to automatic control community, especially for those working in the ILC.

The author expresses his thanks to friends and co-workers from Department of Control Systems and Mechatronics and Department of Computer Engineering, Wrocław University of Science and Technology.

# Chapter 2
# Basic Notions and Notations

This chapter aims to introduce basic notions and the relationships between them. In particular, concepts of repetitive processes, decisions and their constraints as well as loss (goal) functions are discussed. Preliminary mathematical descriptions of them are also provided. Bypassing, notational conventions are presented.

## 2.1 Repetitive Processes

The term *process* has many meanings. In this book it is considered as a synonym or as a word with a close sense to well-established terms such as:

- a dynamical system or cybernetic system,
- a production process,
- an economic or a social process,
- a process of running computations that can be considered broadly, e.g. as running one computer code or as parallel processes governed by an operating system.

Researchers in the artificial intelligence (AI) community also use the term *environment* with a meaning similar to that mentioned above. An associated term, *stochastic process* is also relevant here. Namely, stochastic processes are frequently used as mathematical descriptions of production systems and economic or social processes. This duality of meanings is also used in this book, in cases where it does not lead to misunderstandings. However, in addition to the statistical description of processes, also so-called deterministic models will be used when one can assume that random disturbances have a negligible impact on a process under consideration.

 
W. Rafajłowicz, *Learning Decision Sequences For Repetitive Processes—Selected Algorithms*, Studies in Systems, Decision and Control 401,
https://doi.org/10.1007/978-3-030-88396-6_2

### 2.1.1 Process States

Of crucial importance in a mathematical model of a process is its state at time $n = 0, 1, \ldots$, say. It is further denoted by $s_n$. Roughly speaking, state $s_n$ should contain the minimal information sufficient to calculate (or to predict, in the probabilistic sense) future states $s_{n+1}, s_{n+2}, \ldots$, provided that the model of the process is known, together with external factors influencing the process. It is further assumed that $s_n$ can be selected in such a way that the knowledge of $s_n$ is sufficient for the prediction, independently of how the process arrived at state $s_n$. Stochastic processes having this property are known as Markov processes (see e.g. [67]).

For many applications it suffices to represent process states $s_n$'s as $d_s$-dimensional column vectors in the Euclidean space $R^{d_s}$, as also done in a large part of this book. Occasionally, $s_n$'s with a finite number values for each component $s_n^{(i)}$, $i = 1, 2, \ldots, d$ are also considered. In such cases the ranges of $s_n^{(i)}$'s will be additionally defined.

In more advanced applications, e.g. those when also spatial variables are of importance, one can consider states $s_n$'s taking values in a certain Hilbert space (see [109]), but we shall not follow this generalization in this book.

In a growing number of applications it is also expedient to consider images as process states. In such cases, process states are denoted as $\mathbf{X}_n$, $n = 0, 1, \ldots$, assuming that gray-level images are represented by $l \times m$ matrices. Their elements take values either in $[0, 1]$ interval or in $\{0, 1, \ldots, 255\}$ set nonnegative integers. Clearly, one can formally vectorize $\mathbf{X}_n$'s by stacking their columns into $l\,m$ vectors, but then one loses neighborhood connections that are usually present in images. Thus, further on the vectorization of images is avoided. Color images can be represented by $3 \times l \times m$ tensors. When an ordered sequence of images of the length $L \geq 1$ is considered, it is further denoted as $\mathbb{X}$.

### 2.1.2 States of Repetitive Processes

Repetitive processes form an important subclass of decision processes. They are intensively studied in control systems literature (see, e.g. [111, 140, 147, 170]). As mentioned in the Introduction, possible applications of repetitive processes are much wider, including computer science, transportation systems, batch production systems, social interactions and many others.

Let $N > 1$ be a finite integer that defines the length (the horizon) of a repetitive process. By definition, a repetitive process is repeated many times in the same or similar circumstances. Each repetition is called a pass or a run or a trial (see e.g. [24, 96]). Passes or runs are further designated by an additional index $k$, $k = 0, 1, \ldots$. Consequently, a state of a repetitive process at time $n$ and pass $k$ is further denoted as $s_n(k)$ (or $\mathbf{X}_n(k)$ for images as states). Thus, when a process is repetitive, then one can expect that for pairs of states one has

$$\varrho(s_n(k),\ s_n(k+1)) \le \rho_{max}, \quad n = 0,\ 1,\ \ldots,\ N \tag{2.1}$$

are not too far from each other, as bounded by $\rho_{max} > 0$, in a specified metric $\varrho$ in $R^{d_s}$. A similar requirement is imposed for $\mathbf{X}_n(k)$'s and $\mathbf{X}_n(k+1)$'s. Notice that the requirement (2.1) is weaker than the periodicity of $s_n(k)$'s sequence, since we do not require $s_0(k) = s_N(k)$, $k = 0,\ 1,\ \ldots$. Furthermore, it is allowed that $s_n(k)$ and $s_n(k+j)$ for $j > \pm 1$ are not close to each other in order to allow for learning between passes (runs, trials).

States being random vectors are denoted as $S_n(k) \in R^{d_s}$. Their probability density functions are denoted by $f_s$. These densities can be conditional, with conditions imposed on the previous state and actions, but only one step earlier, in order to retain the Markov property.

Clearly, repetitiveness of a process imposes also requirements on decision sequences and on a model of the process when it is specified. These topics are discussed in the following sections.

## 2.2 Decision Sequences and Disturbances

A decision influences a process under consideration in such a way that it changes its state. Decisions are also called actions or inputs of a process, depending on the branch of science or applications. Later on, these terms are used (almost) interchangeably.

### 2.2.1 *Univariate Decision Sequences*

In the first chapters of this book simple, univariate decision sequences are considered. The following notation is used for them:

$$\vec{x}(k) = [x_1(k),\ x_2(k), \ldots,\ x_N(k)]^{\mathrm{T}}, \tag{2.2}$$

where $x_n(k) \in R$ stands for $n$th decision in $k$th pass (trial), while $^{\mathrm{T}}$ denotes the transposition. Occasionally, $\vec{x}_k$ instead of $\vec{x}(k)$ is used, when complicated operations on $k$ are not present. As a generic symbol for univariate decision sequences $\vec{x}$ is used.

When decisions are random variables, e.g., they are drawn from a specified distribution, then the following notation is used:

$$\vec{X}(k) = [X_1(k),\ X_2(k), \ldots,\ X_N(k)]^{\mathrm{T}}. \tag{2.3}$$

In such cases, particular realizations of random decision sequences will be denoted as (2.2). A generic random sequence of decisions is further denoted by $\vec{X}$, while its realization by $\vec{x}$.

The $\vec{x}$ and $\vec{X}$ notations are used mainly when decision sequences are relatively short ($N$ is from several to dozens of decisions, say).

In many cases one has to impose constraints on decisions or their sequences. For univariate sequences they have the following form:

$$g(\vec{x}(k)) \leq 0, \quad k = 0, 1, \ldots, \tag{2.4}$$

where $g : R^N \rightarrow R^{d_g}$ is a given vector of functions, while the inequalities in (2.4) are understood as component wise. Unless stated otherwise, only the continuity or the piecewise continuity of the components of $g$ is required.

### 2.2.2 *Multivariable Decision Sequences*

Longer and/or multivariable decision sequences (actions) are represented as $d_a$-dimensional column vectors $a_n(k) \in R^{d_a}$, $n = 0, 1, \ldots, N$, where $k$ is a pass (run, trial) number. Notations $\vec{a} \stackrel{def}{=} [a_1, a_2, \ldots, a_N]$ or $\{a_n\}$ are used as generic for a sequence of decisions when there is no need to specify the pass number.

In some cases, particular decisions $a_n^{(i)}(k)$ that are elements of $a_n(k) \in R^{d_a}$, can take only binary $\{0, 1\}$ or integer values.

Similarly to the above, possible constraints on multivariable sequences can be imposed as follows:

$$h(a_n(k)) \leq 0, \quad n = 0, 1, \ldots, N, \; k = 0, 1, \ldots, \tag{2.5}$$

where $h : R^{d_a} \rightarrow R^{d_h}$ is a given vector of continuous or piecewise continuous functions.

Notice that $g$ and $h$ do not depend on pass number $k$, since it seems to be the most common case. However, constraints along each pass $k$ are more frequent. They can be local, i.e. imposed for each $n$ separately, or global, being, e.g., a sum over $n$ of functions of $a_n(k)$'s. Their particular forms are displayed in the following chapters.

### 2.2.3 *Random Disturbances*

Similarly, when disturbances (perturbations) influence the process at $k$th pass, then they are represented as $d_w$-dimensional column vectors $w_n(k) \in R^{d_w}$, $n = 0, 1, \ldots, N$. If disturbances are random vectors, then the notation $W_n(k) \in R^{d_w}$ is used in order to distinguish them from their particular realizations $w_n(k) \in R^{d_w}$. Later on in this book, mainly independent, identically distributed (i.i.d.) random vectors (r.v.'s) are considered. For simplicity of the exposition, it is assumed that their probability density function (p.d.f.), denoted further by $f_w(w)$, $w \in R^{d_w}$ exists.

The same convention is further used for other random variables, vectors and matrices (images). For example, when a process states are observed with random errors that do not influence states of the process, they will be denoted as $\varepsilon$ with indices, while $f_\varepsilon$ denotes the corresponding p.d.f. Exceptions from this convention are made for the problems of classifying images to one of several classes that are labeled as $1, 2, \ldots$. In such cases p.d.f.'s corresponding to each class are denoted as $f_1, f_2, \ldots$ with vector or matrix arguments.

### 2.2.4 Decision Making and Implementation

In this book the emphasis is put on decision making and learning. A subject that makes decisions remains unspecified and is further called an agent. As an agent it can serve

1. a person or a group of persons,
2. a computer system or network, equipped with specialized software,
3. an artificial neural net, a fuzzy system etc.,
4. specialized digital hardware,

the target groups being cases (2) and (3).

Important features that an agent should have include the following:

- an ability to observe a repetitive process (environment),
- possibilities to store these observations for each pass,
- possibilities of computing sequences of decisions as well as
  - storing them,
  - putting them into actions, i.e. influencing the process,
- an ability to learn sequences of decisions, i.e. modifying them in a desirable way, e.g. so as to reduce losses.

It is further assumed that sequences of decisions are put into action as soon as possible. Implementation details are also outside the scope of this book.

It can happen that the implementation of a decision sequence into a process requires complicated devices, which introduce additional delays or dynamic behavior. In such cases it is expected that a mathematical description of such devices is already included in the mathematical model of a process.

## 2.3 Static Models of Processes and Loss Functions

In this section mathematical models of processes are briefly sketched. Their more detailed descriptions are provided in the following chapters. Simultaneously with classes of models also corresponding loss (costs) functions or other indicators of the quality of decision processes are presented in this section.

### 2.3.1 Model-Based Versus Model-Inspired Approaches

It should be stressed that several approaches proposed in this book are model free. In such cases a model is not specified. Additionally, many other approaches are inspired by a model, which means that a model is specified and used for deriving decision rules as well as for establishing theoretical properties of the learning process, but in a practical implementation a model is no longer used—it is replaced by observations of a real process. In model-inspired cases models are also used for simulation purposes, when observations of a real process are not available or if a learning sequence is too short. In such cases it will be pointed out whether observed or simulated learning sequences are used.

### 2.3.2 Deterministic Static Models

The term *static models* is in common use in many branches of science. In general, it refers to models most frequently described as functions of many variables, that are described in terms of decisions (actions, inputs) and outputs (reactions) of a certain process (phenomenon), without taking into account previous decisions and/or hidden variables (states or memory). Static models neglect also the time passed between decisions (inputs) are applied to a process and its reaction, tacitly assuming that possible transient processes have already attained their steady-states. This time can be as short as microseconds or as long as hours, days or months, e.g. when processes in a society are considered.

In many cases there is no possibility of observing the results of applying particular decisions in sequence $\vec{x}$ on a process at hand. One can only observe the final results, usually termed the outputs (reactions) and denoted by $y \in R^{d_y}, d_y \geq 1$. Well-known examples include production of wafers for semiconductor devices, processes in chemical engineering and many others. In the simplest cases, the static model can be described by function $F : R^N \to R^{d_y}$ as $y = F(\vec{x})$ or as

$$y(k) = F(\vec{x}(k)), \quad \text{or} \quad y_k = F(\vec{x}_k), \quad k = 0,\ 1,\ \ldots \tag{2.6}$$

when it is necessary to specify pass number $k$. The basic requirement imposed on $F$ is its continuity or piecewise continuity in a domain of its definition. In many cases static models are more complicated, being composed from static sub-models of sub-processes. Nevertheless, in such cases a general model of the form (2.6) can be obtained,[1] as it is further assumed.

### 2.3.3 Probabilistic Static Models

When a deterministic description of a static model is not adequate, one can consider its description in the terms of p.d.f.'s of $y$. Such models are later called probabilistic models. Two similar, but essentially different from the mathematical point of view, cases should be distinguished. Namely, the first one is when a decision sequence $\vec{X}$ is a random vector and the second one is when a preassigned, deterministic sequence $\vec{x}$ is applied to a process.

**Stochastic inputs case** Assuming the existence of the conditional p.d.f. of $y$ given $\vec{X} = \vec{x}$, denote it by $f_y(y|\vec{x})$. Then, at pass $k$, the following model can be applied:

$$y_k \text{ is drawn at random according to } f_y(y|\vec{x}_k),\ k = 0,\ 1,\ \ldots \tag{2.7}$$

for $\vec{x}_k$ 's being realizations of $\vec{X}_k$'s, and analogously for $\vec{X}(k)$ etc. notation. The conditional p.d.f. of $y$ is given by

$$f_y(y|\vec{x}) = \frac{f_{\vec{x}y}(\vec{x},\ y)}{f_{\vec{x}}(\vec{x})}, \tag{2.8}$$

where $f_{\vec{x}y}(\vec{x},\ y)$ is the joint p.d.f. of $y$ and $\vec{x}$, while $f_{\vec{x}}$ stands for the marginal p.d.f. of $\vec{X}$, i.e.,

$$f_{\vec{x}}(\vec{x}) = \int_{R^N} f_{\vec{x}y}(\vec{x},\ y)\, dy. \tag{2.9}$$

The estimation of $f_y(y|\vec{x})$ is usually based on (2.8), but it requires a large number of observations. When they are not available, a simplified regression model of the following form is used

$$\mu(\vec{x}) = \mathbb{E}_y\left(y|\vec{X} = \vec{x}\right) = \int_{R^{d_y}} y\, f_y(y|\vec{x})\, dy, \tag{2.10}$$

where $\mathbb{E}_y\left(y|\vec{X} = \vec{x}\right)$ denotes the conditional expectation of $y$, given $\vec{x}$ (see [53, 55] for estimating such models).

**Deterministic decisions case** A sequence of decisions $\vec{x}$ frequently arises as a result of human, more or less rational, decisions. It can also be generated by a computer as the result of applying a certain policy. In both cases it is rather difficult

---

[1] When a structure of sub-models is complicated, then it may not be easy to obtain the description (2.6) in a simple, closed form, but for the purposes of this book it suffices to ensure that it is—at least conceptually—possible.

to interpret $\vec{x}$ as a random vector. Therefore, it is further called a deterministic decision sequence.

Denote by $f_y(y;\ \vec{x})$ a p.d.f. of $y$ that depends on a deterministic decision sequence $\vec{x}$. From the mathematical point of view, $\vec{x}$ is interpreted as a vector of parameters, influencing the probability distribution of $y$. In order to distinguish this case from the stochastic one, a semi-colon is used to separate $y$ and $\vec{x}$. With this notation, the following model can be applied at pass $k$ for selected $\vec{x}_k$: for $k = 0,\ 1,\ \ldots$

$$y_k \text{ is drawn at random according to } f_y(y;\ \vec{x}_k). \tag{2.11}$$

The following, frequently-used class of models illustrates the (2.11) scheme. Suppose that in (2.6) $y_k$'s are observed with additive, independent, identically distributed random errors $W_k$ with realizations $w_k \in \mathrm{R}^{d_w}$'s, $d_w = d_y$ and having common p.d.f. $f_w(w)$, i.e.

$$y_k = F(\vec{x}_k) + w_k, \quad k = 0,\ 1,\ \ldots . \tag{2.12}$$

In such cases it will always be assumed that random errors have a zero mean and a finite covariance matrix.

It is easy to verify that under the above assumptions one obtains:

$$f_y(y;\ \vec{x}_k) \ = \ f_w\left(y - F(\vec{x}_k)\right), \quad k = 0,\ 1,\ \ldots . \tag{2.13}$$

Thus, the stochastic model (2.11) simplifies to (2.12).

Returning to a general class of stochastic models with deterministic decision sequences, it may happen that the number of available observations is not sufficient for their nonparametric estimation. In such cases one may use the regression function that is defined as follows:

$$m(\vec{x}) \ = \int_{\mathrm{R}^{d_y}} y\, f_y(y;\ \vec{x})\, dy\,. \tag{2.14}$$

In particular, in a special case (2.12), one obtains: $m(\vec{x}) = F(\vec{x})$, as expected.

**Remark 2.1** (*Stochastic versus deterministic decision sequences*) The main difference between stochastic models with stochastic or deterministic decision sequences is that the latter formula (2.8) does not have its counterpart. This fact implies that the methods of estimating $f_y(y|\ \vec{x}_k)$ and $f_y(y;\ \vec{x}_k)$ are different. (see [137] for duscussion on this topic)

### 2.3.4 *Assessing the Quality of Decision Sequences*

The terminology of naming the method of assessing the quality of decisions is quite different, depending on the branch of science, assumptions made, etc. The mostly commonly used terms are the following: cost, loss or goal function, regret, quality criterion or index and many others. Later on, they are considered as synonyms. Their usage in this book is context dependent, taking into account habits in a particular sub-discipline.

#### 2.3.4.1 Goal Functions for Deterministic Models

For static models it is customary to define cost function as follows: $c : R^N \times R^{d_y} \to R$ is a real-valued (most frequently nonnegative) function that attaches cost $c(\vec{x},\ y)$ to decision sequence $\vec{x}$, which—when applied—invokes the process response $y$, according to the model $y = F(\vec{x})$. It is widespread to define goal function $J$ by substituting $y$ into $c$, i.e.,

$$J(\vec{x}) = c(\vec{x},\ F(\vec{x})), \tag{2.15}$$

even if this substitution is not easy to perform in practice.

It is also typical to assume that $J(\vec{x})$ is differentiable with respect to $\vec{x}$ or at least continuous[2] in a domain of interest $\mathcal{X} \subset R^N$, say, as is made more precisely in subsequent chapters. Staying within the classical optimization framework, it is assumed that $\mathcal{X}$ is a close and bounded subset of $R^N$. These assumptions imply that $\mathcal{X}$ is a compact set, which—together with the continuity of $J$ on $\mathcal{X}$—yield that the following optimization problems have at least one solution: find $J^*$ and $\vec{x}^*$ such that

$$J^* = \min_{\vec{x} \in \mathcal{X}} J(\vec{x}), \qquad \vec{x}^* = \arg\min_{\vec{x} \in \mathcal{X}} J(\vec{x}). \tag{2.16}$$

In the first case the emphasis is put on the lowest possible value of the goal function, while in the second one the minimizing sequence of decisions is of more interest. In both cases finding the minimum is understood as the global minimum of $J$ over $\mathcal{X}$.

Optimization problems with more than one criterion function are also intensively studied in the literature. In this book, the emphasis is put on problems with one criterion, as sufficiently difficult when the number of decisions $N$ is large. However, in the following chapters bi-criteria problems are considered as a possible way of taking constraints into account.

[2] Optimization problems with non-differentiable or discontinuous functions $J$ are also considered (see, e.g. [70, 161]).

#### 2.3.4.2 Quality Criteria for Probabilistic Models

When probabilistic models with deterministic decision sequences are considered, their quality can be assessed by the following goal function

$$J(\vec{x}) = \int_{\mathrm{R}^{d_y}} c(\vec{x},\, y)\, f_y(y;\, \vec{x})\, dy\,, \tag{2.17}$$

where $c$ is a cost function such that the above integral is finite and defines a continuous function for $\vec{x} \in \mathcal{X}$. The integral on the right-hand side of (2.17) is further denoted as $\mathbb{E}_{Y(\vec{x})}\,[c(\vec{x},\, Y(\vec{x}));\, \vec{x}]$ where $Y(\vec{x})$ has p.d.f. $f_y(y;\, \vec{x})$, while the semi-colon notation serves to indicate that $\vec{x}$ is temporarily fixed and interpreted as parameters of the probability density function.

**Quadratic criterion** Consider an important special case of a static model (2.12) with scalar output ($d_y = 1$) and random errors with zero mean and finite variance $\sigma_w^2 > 0$. and cost function

$$c(\vec{x},\, y) = \psi_0(\vec{x}) + \psi_1(\vec{x})\,(y^* - y)^2\,, \tag{2.18}$$

where $\psi_0(\vec{x})$ and $\psi_1(\vec{x})$ are given, scalar valued nonnegative functions that do not depend on $y$, while $y^*$ is a preselected target of a process response (output). Then, by (2.13),

$$J(\vec{x}) = \mathbb{E}_{Y(\vec{x})}\,[c(\vec{x},\, Y(\vec{x}));\, \vec{x}] = \tag{2.19}$$

$$= \psi_0(\vec{x}) + \psi_1(\vec{x}) \int_{\mathrm{R}^{d_y}} (y^* - y)^2\, f_w(y - F(\vec{x}))\, dy\,,$$

which yields

$$J(\vec{x}) = \psi_0(\vec{x}) + \psi_1(\vec{x})\,(y^* - F(\vec{x}))^2 + \sigma_w^2\, \psi_1(\vec{x}). \tag{2.20}$$

Observe that in (2.20) $J(\vec{x})$ depends on random errors through their variance $\sigma_w^2$ only.

**Min-max and max-max criteria** In this and similar special cases the full knowledge of $f_y(y;\, \vec{x})$ is not required. However, for many other, important in practice quality criterions, one has to know or to estimate $f_y(y;\, \vec{x})$. The most explicit example is the following goal function for single response processes:

$$J(\vec{x}) = \max_{y \in \mathrm{R}} f_y(y;\, \vec{x})\,, \tag{2.21}$$

where for multi-modal p.d.f.'s the maximum is understood as the global one. Denote by $y_{mod}$ (dependent on $\vec{x}$ ) the mode of $f_y(y;\, \vec{x})$. For sufficiently regular p.d.f.'s and $\epsilon > 0$ small enough, the expression

$$\int_{y_{mod}-\epsilon}^{y_{mod}+\epsilon} f_y(y;\ \vec{x})\,dy \tag{2.22}$$

is the probability that, for given $\vec{x}$, the process response $y$ is near its highly probable value $y_{mod}$.

At least two cases are worth highlighting.

**Case 1** Keeping the process output near $y_{mod}$ is desirable for its proper functioning. Then, the goal is to find $\vec{x}^{**}$ for which

$$\vec{x}^{**} = \arg\max_{\vec{x}\in\mathcal{X}} \left[ \max_{y\in R} f_y(y;\ \vec{x}) \right]. \tag{2.23}$$

**Case 2** The most frequently appearing responses near $y_{mod}$ are unsatisfactory process behaviors. In such cases the goal is to find $\vec{x}^{*}$ for which

$$\vec{x}^{*} = \arg\min_{\vec{x}\in\mathcal{X}} \left[ \max_{y\in R} f_y(y;\ \vec{x}) \right]. \tag{2.24}$$

Examples of possible applications of Case 1 include decision making in statistical process control (SPC), where it is desirable to keep a process output close to a target value $y_{mod}$, but also attempt to reduce its variability around it (see, e.g. [95, 99]) by selecting decision sequence $\vec{x}$. Notice, however, that in SPC it is typical to assume that $f_y$ is the p.d.f. of the Gaussian distribution due to an insufficient number of observations. Nowadays, more and more frequently a large number of measurements is available which allows to estimate $f_y$ in a nonparametric way [137] and then use (2.23).

### 2.3.5 *Illustrative Example*

To illustrate possible applications of Case 2, consider Fig. 2.1—the solid line. It is a log-normal distribution with the mean $\mu = 0$ and the standard deviation $\sigma = 0.25$. This p.d.f. can be used as a simple model for maintaining the distance between a group of people talking. Such behavior is proper in typical cases. However, in the context of the COVID-19 pandemic it is desirable to maintain greater distances, e.g. by decisions that lead to flat distributions of social distances, as illustrated in Fig. 2.1 by the dashed and dotted lines, which were obtained by setting $\mu = 0.5$ and $\mu = 1$, respectively. Thus, one can select $\vec{x} = \mu$ as a decision variable[3] that influences $f_y$, which—for the log-normal distribution—has the form:

[3] In this simple example the decision sequence has only one element.

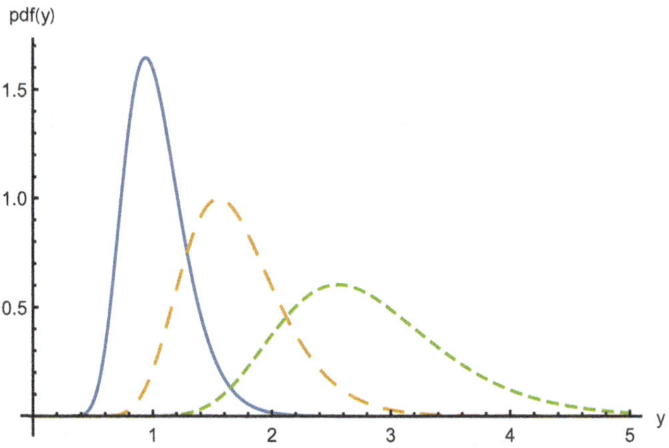


**Fig. 2.1** An illustration for Case 2—influencing to maintain greater social distances when society is exposed to a pandemic threat, such as COVID-19

$$f_y(y;\bar{x}) = \begin{cases} \dfrac{e^{-\frac{(\log(y)-\bar{x})^2}{2\sigma^2}}}{\sqrt{2\pi}\sigma\, y} & y > 0\,, \\ 0 & y \le 0\,. \end{cases} \tag{2.25}$$

According to (2.21), one has to find the argument $y^*$ of the maximum of (2.25) with respect to $y$, considering $\vec{x}$ as a parameter, which yields

$$y^*(\vec{x}) = \exp[\vec{x} - \sigma^2]\,, \quad \vec{x} > 0\,. \tag{2.26}$$

By substituting (2.26) into (2.25) one obtains:

$$J(\vec{x}) = \frac{\exp\left[\frac{\sigma^2}{2} - \vec{x}\right]}{\sqrt{2\pi}\sigma}\,, \quad \vec{x} > 0\,. \tag{2.27}$$

In practice, however, the distance between people can not be too large, otherwise they would not be able to hear each other. As is known, a sound's loudness decreases with the squared distance between speakers. This fact leads to the following constraint $(\vec{x})^{-2} \ge \zeta^{-1}$, or equivalently,

$$(\vec{x})^2 \le \zeta, \quad \vec{x} > 0\,, \tag{2.28}$$

where $\zeta > 0$ is a prescribed threshold.

Assume that $\sigma > 0$ is independent of $\vec{x}$ . Then, due to the strict monotonicity of the log(.) function, it is convenient to minimize the logarithm of $J$, which leads to the following Lagrange function:

**Fig. 2.2** The Lagrange function (2.29) in Example 2.3.5

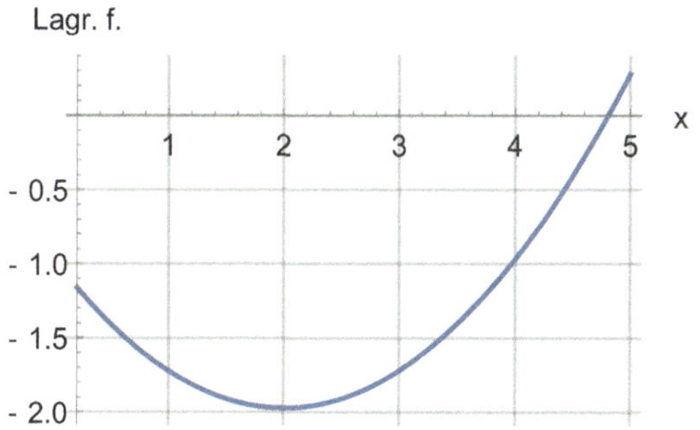


$$L(\vec{x}, \lambda) = \left[\frac{\sigma^2}{2} - \vec{x}\right] + \lambda\,((\vec{x})^2 - \zeta) \tag{2.29}$$

that is minimized by $\vec{x}^*(\lambda) = 1/(2\,\lambda)$, which meets the constraint (2.28) for $\lambda = 1/(2\,\sqrt{\zeta})$ . Thus, $\vec{x}^* = \sqrt{\zeta}$ that for $\zeta = 4$ provides quite a reasonable social distance of 2 meters. The plot of (2.29) function for $\lambda = 1/(2\,\sqrt{\zeta}) = 0.25$ and $\sigma = 0.25$ is shown in Fig. 2.2.

## 2.4 Models of Dynamic Processes

In many applications static models of processes are not sufficient and one has to take into account their transient behavior. Furthermore, for certain processes a static model does not formally exist. The simplest examples are processes that integrate inputs and their discrete time counterparts. Roughly speaking, dynamic processes can be characterized by the fact that their responses (reactions) depend not only on present actions, but also on actions (inputs) undertaken in the past. Sometimes it is formulated differently that such processes have memory.

In this section mathematical descriptions of dynamic processes are briefly presented, confining them to those processes that can be described in terms of finite dimensional states (see Sect. 2.1). It is assumed that states are defined in such a way that the knowledge of a present state implicitly contains all the information about previous process behaviors, independently of a way which led to this state. Briefly, the emphasis is on dynamic processes having the Markov property.

### *2.4.1 Markov Chains Models*

A repetitive Markov process with discrete time $n = 0, 1, \ldots, N$ and trials (passes) indexed by $k = 0, 1, \ldots$ can be described as follows. For pass $k$ draw at random an initial state $s_0(k) \in R^{d_s}$ as a random vector having p.d.f. $f_{s0}(s)$. Then, for $n = 1, 2, \ldots, N$ repeat the following steps:

$$\text{draw } s_n(k) \text{ according to } f_{sn}(s_n(k)|s_{n-1}(k),\ a_n(k)), \tag{2.30}$$

where $f_{sn}(s_n(k)|s_{n-1}(k),\ a_n(k))$ is the conditional p.d.f. of $S_n(k)$ given previous state $s_{n-1}(k)$ and action $a_n(k)$). $f_{sn}$ is called the transition density of a controlled Markov chain. In many cases it suffices to consider stationary Markov chains for which transition densities $f_{sn}$ do not depend on $n$ and is further denoted by $f_s$. If, additionally, actions are deterministic, then transition densities are denoted as $f_s(s_n(k)|s_{n-1}(k);\ a_n(k))$. Under these assumptions, nonparametric problems of estimating $f_s$ become more realistic, when a sufficient number of passes of a repetitive process is observed and stored.

An important sub-class of repetitive, discrete time Markov chains models has the following form: for $n = 0, 1, \ldots, N$, and $k = 0, 1, \ldots$

$$s_{n+1}(k) = G(s_n(k),\ a_n(k)) + w_n(k), \tag{2.31}$$

where $s_0(k)$, $k = 0, 1, \ldots$ are given and frequently they are the same or have similar values for all $k$. In (2.31) $G : R^{d_s} \times R^{d_a} \to R^{d_s}$ is a given continuous mapping, while $w_n(k)$'s are i.i.d. realizations that are drawn according to $f_w(w)$, which is the p.d.f. of $d_s$-dimensional random vectors, having a zero mean vector and a finite covariance matrix.

Under these assumptions, (2.31) generates a Markov process with the transition density of the form:

$$f_s(s_n(k)|s_{n-1}(k),\ a_n(k)) = f_w\left(s_n(k) - G(s_{n-1}(k),\ a_n(k))\right), \tag{2.32}$$

for $n = 0, 1, \ldots, N$, $k = 0, 1, \ldots$.

### 2.4.2 Deterministic Models and Miscellaneous Remarks

If in (2.31) variances of elements of random errors $w_n(k)$ are zero (or very close to zero), one can use the following deterministic models of dynamic repetitive processes: for $n = 0, 1, \ldots, N$, and $k = 0, 1, \ldots$

$$s_{n+1}(k) = G(s_n(k),\ a_n(k)), \tag{2.33}$$

where $s_0(k)$, $k = 0, 1, \ldots$ are given.

When univariate decision sequences are considered, then the description:

$$s_{n+1}(k) = G(s_n(k),\ x_n(k)) \tag{2.34}$$

is used for dynamic models with obvious changes when probabilistic models are considered.

**Remark 2.2** In the previous sections the following notations were introduced for univariate decision sequences: $\vec{x}$ for a generic sequence of length $N$ and $\vec{x}(k)$ or $\vec{x}_k$ for a sequence applied at $k$th pass. In parallel, for multivariable decision sequences:

- $\{a_n\}_{n=0}^{N}$ or $\{a_n\}$ or $\vec{a}$ stand for a generic sequence of decisions, where $\vec{a}$ is the $d_a \times (N+1)$ matrix having $a_n$'s as its columns,
- $\{a_n(k)\}$ or $\vec{a}(k)$ have the same meaning as above, when $k$th pass (trial) is considered.

**Remark 2.3** In many cases not all states of a process can be observed and one is forced to make decisions from observations of the process outputs (responses) of the form: for $n = 0,\ 1,\ \ldots,\ N$, and $k = 0,\ 1,\ \ldots$

$$y_n(k) \;=\; O(s_n(k)) \;+\; (\varepsilon_n(k)) \tag{2.35}$$

where $y_n(k) \in \mathrm{R}^{d_y}$ are output vectors that result from the process states $s_n(k)$ by mapping a continuous $O\,:\,\mathrm{R}^{d_s} \times \mathrm{R}^{d_a} \rightarrow \mathrm{R}^{d_y}$. This mapping may be as simple as the selection of elements of $s_n(k)$ that can directly be observed or more complicated, when elements of $s_n(k)$ are combined in a possibly nonlinear way.

Additionally, random errors $\varepsilon_n(k)$'s can corrupt output observations. In such cases, it is assumed that they have a zero mean vector and a finite covariance matrix.

**Remark 2.4** Model (2.33) and the previously described models are collections of repetitions of runs (passes) of the same model that are not linked to each other. In the following chapters, however, these passes are linked by decision learning procedures. For example, sequence of actions $a_n(k)$, $n = 0,\ 1,\ \ldots,\ N$ is computed on the basis of previous actions $a_n(k-1)$, $n = 0,\ 1,\ \ldots,\ N$ and observations of previous process states $s_n(k-1)$, $n = 0,\ 1,\ \ldots,\ N$ as well as on the basis of actions $a_n(k)$, $n = 0,\ 1,\ \ldots,\ j$ and states $s_n(k)$, $n = 0,\ 1,\ \ldots,\ j$ of the current pass $k$, but only up to its local time $j < N-1$, which is the last one available at present.

These links between passes makes the considered process to be a 2D process in the sense that one has to consider two independent variables $n$ and $k$ for multidimensional process states $s_n(k)$ and actions $a_n(k)$.

Notice that one is faced with 2D processes also when their static models are considered, since in searching for the optimal decision sequence that minimizes $J(\vec{x})$ it is customary to improve $\vec{x}_k$ on the basis of $\vec{x}_{k-1}$ and $J(\vec{x}_{k-1})$, but $\vec{x}_{k-i}$, $i = 1,\ 2,\ \ldots,\ k$ are univariate decision sequences themselves.

### 2.4.3 Quality Criteria for Dynamic Processes

The goal functions discussed in Sect. 2.3.4 can also be adopted as decision quality criteria for dynamic processes, e.g. by replacing the output of a static process by $s_N(k)$ or $y_N(k)$. However, it frequently happens that decision sequences for dynamic processes are longer and there are possibilities of attaching partial losses to each decision and then summing them up. Such a property of quality criterion is of fundamental importance because one may try to split searching for optimal decisions into a number of interconnected sub-problems of smaller dimensionality.

The most widely-used decision quality criterion is of the form:

$$J(\vec{a}) = \sum_{n=1}^{N} \phi_n(s_n, a_{n-1}), \tag{2.36}$$

where $\phi_n : \mathrm{R}^{d_s} \times \mathrm{R}^{d_a} \to \mathrm{R}$, $n = 1, 2, \ldots, N$ is a sequence of given functions, while $s_n$'s and $a_{n-1}$'s are linked by a model of a process:

$$s_{n+1} = G(s_n, a_n), \quad n = 0, 1, \ldots, (N-1) \tag{2.37}$$

where $s_0$ is given. Frequently, functions $\phi_n$ do not depend on $n$ and notation $\phi : \mathrm{R}^{d_s} \times \mathrm{R}^{d_a} \to \mathrm{R}$ is used or

$$\phi_n(s_n, a_{n-1}) = \gamma^n \phi(s_n, a_{n-1}), \quad n = 0, 1, \ldots, (N-1), \tag{2.38}$$

where $0 < \gamma < 1$ is called the discount rate. It is however noteworthy that the selection of $\gamma > 1$ may also be of interest when one wants to force a desired process behavior at earlier decision stages.

The problem of the minimization of (2.36) and (2.37) with respect to $\vec{a}$ is frequently formulated as a problem with additional constraints, imposed on $\vec{a}$, e.g.

$$h(a_n) \le 0, \quad n = 0, 1, \ldots, (N-1). \tag{2.39}$$

Constraints can also be imposed on the process states.

For univariate decision sequences (2.36) reduces to

$$J(\vec{x}) = \sum_{n=1}^{N} \phi_n(s_n, x_{n-1}). \tag{2.40}$$

For stochastic models (2.30) and (2.31) the most frequently used counterpart of (2.36) has the form:

$$J(\vec{a}) = \mathbb{E}\left[\sum_{n=1}^{N} \phi_n(s_n, a_{n-1})\right], \tag{2.41}$$

where the expectation is taken with respect to all states $s_n$'s and also to all actions, if they are random vectors. Additional constraints, analogous to (2.39), can also be imposed on random actions, e.g.,

$$\mathbb{E}\left[h(a_n)\right] \leq 0, \quad n = 0,\ 1,\ \ldots,\ (N-1) \tag{2.42}$$

as well as on states $s_n$.

# Chapter 3
# Learning Decision Sequences and Policies for Repetitive Processes

Learning is an inherent ability common to human beings[1] and animals.

Learning in a broad sense is a very complicated phenomenon, involving physiological, social and psychological aspects, among others. These aspects are far beyond the scope of this book.

The main emphasis is on artificial learning algorithms that are dedicated for implementation in computer systems or other specialized digital devices.

The reader is referred to [85] for a discussion and the state of the art on current learning abilities of human beings and algorithms.

## 3.1 Learning Decisions and Decision Sequences

The aim of considered learning algorithms is to improve decisions (actions) made by such devices or to support decisions made by humans.

However, it is worth emphasising the features that are common for biological and artificial learning algorithms, namely,

1. a need for interactions with a process (an environment), which provide feedback information on whether the learning process leads to more satisfactory decisions, actions or behaviors,
2. the presence of a kind of memory (storage space) that is necessary to collect learning examples, at least more recent ones,
3. the possibility of repeating and correcting decisions (actions).

In this book, the emphasis is put on the last feature, since it is even more important when learning a whole decision sequence instead of separate decisions.

[1] Recently, learning abilities of plants, e.g. trees, have also been investigated.

 
W. Rafajłowicz, *Learning Decision Sequences For Repetitive Processes—Selected Algorithms*, Studies in Systems, Decision and Control 401,
https://doi.org/10.1007/978-3-030-88396-6_3

### 3.1.1 Remarks on Learning in Control Systems

It seems that one of the first methods that were able to learn decisions was the extremum seeking device proposed by Leblanc in 1922 (see [155] for a description of this ingenious idea and for the bibliography of this—still active—area of research and applications). In its original statement, this method was developed for seeking a local extremum of an unknown function by using the cosine function as a dither signal. The phase of the response signal was used to determine whether to continue searching to the left or to the right from the present position or to stay temporarily in the current position. It is also worth mentioning that the Leblanc's method has the following features:

- it is model-free in the sense that the function to be optimized is unknown,
- periodicity invoked by the cosine excitation,
- it has the ability to learn new positions of an extremum when the extremum is slowly drifting, since the dithering signal is supplied permanently.

Over the last one hundred years, many other learning algorithms were developed in control theory. This stream of research is known as adaptive control algorithms (see, e.g. for recent monographs [174] and the bibliography cited therein). A common feature of many of them is that only one decision at each instant of time is considered.

However, there are at least three important classes of control tasks that takes sequence of decisions into account. These are:

1. optimal control problems,
2. iterative learning control,
3. run-to-run control approaches.

Many common features with the tasks mentioned above also have the stream of research known as reinforcement learning.

### 3.1.2 Selected Learning Tasks in Operations Research

The need for learning decision sequences considered as whole entities emerged also in many other branches of science and engineering. The following list contains only the most widespread classes of such tasks that emerged in operations research.

1. Playing chess and other games that consist of a sequence of movement of pieces has been a challenge for learning algorithms for more than sixty years. In [45] the algorithm for self-learning, evolutionary chess playing, which achieved significant success, is described. In the same paper one can find also a history and the bibliography of learning approaches to chess playing, including those that are based on reinforcement learning. The idea of using the differential evolution algorithm for tuning a chess evaluation function is presented in [21]. The present

state of the art in playing chess, Go and other computer games can be found in [38].

2. The well-known traveling salesman problem (TSP) is stated as the task of finding the shortest route of visiting all the nodes (e.g. cities) in a given graph with distances between them and adhering to the following rules: each node is visited only once, with the exception of the first node (a base), which should be the beginning and the end of the route (see, e.g. [61, 154] for basic facts and methods of attacking this problem). From the formal point of view there is nothing to learn in this precisely defined problem. However, as the optimization problem TSP is classified as the NP-hard one. Thus, when the number of nodes is large and connections among them are dense, one has to use advanced heuristic algorithms (see [164]), including those that are based on learning (see [9, 23, 35, 125] for an excerpt of recent contributions).
   Also variants of TSP are considered in which distances between the nodes can change over time. They are named Dynamic TSP (DTSP). Depending on a precise problem statement, for DTSP's learning is even more natural than for original TSP's (see [54, 91, 162, 163] for applications and recent approaches).
3. Problems of scheduling operations (tasks) have many features in common with TSP's. In particular, one has to consider scheduling whole sequences of decisions, which are—in general—NP-hard optimization problems. For the same reasons as above, a large number of meta-heuristic algorithms was developed.
   In the dynamic scheduling of tasks, the role of learning is even more apparent due to a larger number of uncertainty sources, including execution times, adding or removing jobs, failures of machines and many others. The reader is referred to [127, 160, 177] for learning-based scheduling directed to manufacturing systems. An interesting area of applications of learning for dynamic scheduling can be found in [142, 159], where learning is used for software engineering tasks.

Learning sequences of decisions play an increasingly important role in cyber-security systems. These topics are not discussed here, since most of the approaches are related to machine-learning methods.

The dichotomy: learning separate decisions versus learning whole decision sequences is less sharp when learning decision rules (policies) is considered. This issue is addressed later.

## 3.2 How Algorithms Can Learn

In this section a brief review of learning algorithms is given with the emphasis on what kind of results they are able to produce. The other aspect discussed here includes approaches to constructing such algorithms, taking into account the roles played by data, interactions with a process (an environment) and their mathematical models.

### 3.2.1 *Learning Directly from a Static Process—Disturbance Free Case*

To fix ideas, consider the simplest case, described in Sect. 2.3, assuming that

1. for every sequence of decisions $\vec{x}(k)$, output $y(k)$ of the process $y = F(\vec{x})$ is measurable without essential errors,
2. the goal function (process quality criterion) $J(\vec{x})$ is known and can be sufficiently accurately computed as follows:

$$J(\vec{x}(k)) = c(\vec{x}(k),\; F(\vec{x}(k))), \tag{3.1}$$

3. possible constraints on admissible $\vec{x}(k)$ are given.

It should be stressed that in the considered case $F$ remains unknown, and one can observe only $y(k) = F(\vec{x}(k))$.

#### 3.2.1.1 Disturbance Free Case

The goal of learning is finding decision sequence $\vec{x}^*$ for which $\min_{\vec{x}} J(\vec{x})$ is attained. The above problem statement suggests that one can consider many known optimization methods as learning algorithms.

A sufficiently general class of learning algorithms can be described as follows: for $k = 0,\ 1,\ \ldots$

$$\vec{x}(k+1) \;=\; \Phi_k\left((\vec{x}(k),\ J(\vec{x}(k))),\ \ldots,\ (\vec{x}(0),\ J(\vec{x}(0)))\right), \tag{3.2}$$

where $\Phi_k$, $k = 0,\ 1,\ \ldots$ is a sequence of mappings with a growing number of arguments that produces sequences of decisions to be used in the following passes. In general, it is allowed that $\Phi_k$'s are random mappings. $\vec{x}(0)$ is a starting point that can be based on previous experience or selected at random.

**Remark 3.5** Some optimization methods, e.g. evolutionary ones, do not fit to (3.2), since they produce more than one $\vec{x}$ in each iteration. Observe, however, that if one wants to apply such methods as learning schemes that interact with a process, then each $\vec{x}$ generated has to be fed as the process input separately (see Fig. 3.1). Hence, also evolutionary algorithms can be described by (3.2).

Learning scheme (3.2) is too general for more detailed analysis. For this reason particular classes of them are considered. Many of them have the following form, which is typical for the majority of learning algorithms,

$$\vec{x}(k+1) \;=\; \vec{x}(k) \;+\; \gamma_k\, \vec{d}(k), \quad k = 0,\ 1,\ \ldots \tag{3.3}$$

where $\vec{d}(k) \in \mathrm{R}^N$ is a direction of search for a better decision sequence, while $\gamma_k$ is a step made in direction $\vec{d}(k)$.

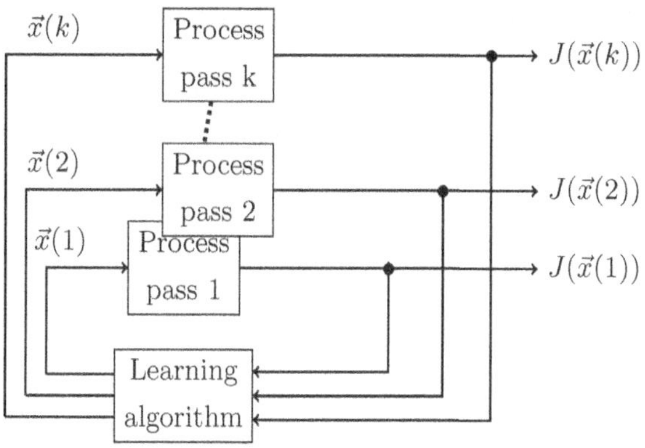


**Fig. 3.1** An outline of a repetitive learning algorithm interacting with a static process

In general, $\vec{d}(k)$ may depend on all previous pairs $(\vec{x}(k),\ J(\vec{x}(k))$, but in practice the results of only a small number of previous passes are stored and used in computing $\vec{d}(k)$. Clearly, (3.3) is a special case of (3.2) by selecting

$$\Phi_k\left((\vec{x}(k),\ J(\vec{x}(k))),\ \ldots,\ (\vec{x}(0),\ J(\vec{x}(0)))\right) = \qquad (3.4)$$

$$= \vec{x}(k) + \gamma_k\,\vec{d}(k).$$

The first candidate for $\vec{d}(k)$ would be

$$\vec{d}(k) = -\text{grad}_{\vec{x}}\,J(\vec{x})\Big|_{\vec{x}=\vec{x}(k)}. \qquad (3.5)$$

However, this choice is not admissible, since $J$ is unknown. One can only approximate the gradient in (3.5) from observations of $J$ (see the next subsection).

In model-based optimization algorithms $\gamma_k$'s are frequently selected by a line search along $\vec{d}(k)$, which requires a large number of trials. In on-line versions an additional number of passes is frequently too expensive and $\gamma_k$'s are selected either as a small constant or as a deterministic sequence slowly decaying to zero. An adaptive choice, but based solely on pairs $(\vec{x}(k),\ J(\vec{x}(k))$ that are known from previous passes, is also possible.

A closer look at (3.3) reveals subtle differences between an algorithm that learns from a process (environment) and a similar optimization algorithm. The main difference is in the necessity of interacting with a repetitive process when running (3.3), as it is emphasised in its description that follows, while the underlying optimization algorithm relies only on computations of $F$ and $J$.

#### 3.2.1.2 A Skeletal Version of the Learning Algorithm for Static Processes

**Step 0** Preparations: select the rule for calculating $\vec{d}(k)$'s and $\gamma_k$'s. Select starting sequence $\vec{x}(0)$ and set $k = 0$.
**Step 1** Apply sequence of decisions $\vec{x}(k)$ to the process (environment).
**Step 2** When $k$th pass is finished, acquire the process response $y(k)$, then compute and store $J(\vec{x}(k))$.
**Step 3** Compute the update

$$\vec{x}(k+1) = \Phi_k\left((\vec{x}(k),\ J(\vec{x}(k))),\ \ldots,\ (\vec{x}(0),\ J(\vec{x}(0)))\right) \tag{3.6}$$

or—in simpler cases –

$$\vec{x}(k+1) = \vec{x}(k) + \gamma_k\, \vec{d}(k) \tag{3.7}$$

**Step 4** Set pass number $k := k + 1$, store $\vec{x}(k)$ and go to Step 1.

Interactions of a learning algorithm with a process (environment) that are pointed out in Step 1 and Step 2 are also displayed in Fig. 3.1. Implementation details of these interactions can be very different and they are outside the scope of this book. It is, however, worth noticing that in addition to classic sensors and actuators one can apply also much richer sources of information about a process such as images, 3D images and image sequences.

The second difference between this algorithm and the underlying optimization algorithm is in the possible requirement of working in real-time. Although the optimized process is static, it can be so fast that its repetitions impose a hard limit constraint on the time left for executing Step 3.

This, in turn, excludes from considerations many advanced optimization methods as possible candidates for on-line learning algorithms for longer decision sequences or significantly limits their applicability [14]. In particular, optimization methods based on second derivatives of $J$ are difficult to apply, unless one has a model of a process supporting computations (see Chap. 5 for an illustrative example).

If, additionally, the underlying optimization method is expected to search for the global minimum, then good candidates are: the classic Nelder–Mead method (see [84] and the bibliography cited therein) and the differential evolution approach (see also Chap. 4). Their common feature is an implicit way of using the gradient of $J$ (see [20]), which—additionally—is cheaply evaluated by one additional trial at each iteration (see also [136]). This feature, plus a certain degree of robustness against perturbations in observing $J$ (see [20] and Chap. 5) make them attractive as learning algorithms.

#### 3.2.1.3 Convergence of the Learning Algorithm

An important factor in selecting a learning algorithm is its convergence to the global minimum of $J$. When classic optimization methods are considered as learning algorithms, then sufficient conditions for their convergence, under more or less strengthened assumptions are well known (see e.g. [152]) Sufficient conditions for the convergence of the Nelder–Mead algorithm can be found in [84].

The topic of convergence of the differential evolution learning algorithm is more complicated due to a randomness that is built-in into its construction. For this reason, one has to consider its convergence in a probabilistic sense, e.g. with the probability one or convergence in probability. The latter mode of convergence is weaker, but more frequently discussed as easier to investigate.

As is known (see [72]), the original version of the differential evolution algorithm may not be convergent, even in probability. However, for its modifications a global convergence in probability can be proved [72]. The reader is referred to [50, 59, 64] for discussions of various aspects of their convergence.

There were also several modifications of differential evolution algorithms dedicated to either multi-objective optimization or to incorporating constraints into the search procedure (see [93, 176, 186]). In [133] a differential evolution algorithm, which is able to handle constraints, is proposed and tested. It is based on a bi-objective population filtering search. In [135] it was applied to a rather difficult problem of searching for optimal control for processes which are described by algebraic-integral equations. In Chap. 4 the idea of a bi-objective differential evolution with population filtering is further developed and extensively tested on the COVID-19 decision-making example (see Chap. 5).

### *3.2.2 Learning Directly from a Static Process—Observations with Random Errors*

As mentioned in the previous subsection, the differential evolution and the Nelder–Mead method can serve as learning algorithms when random errors in observing $J$ or the process output are present. However, it is worth briefly discussing other methods that were originally designed to cover such cases and to assess to what extent they might be useful as underlying methods for learning algorithms, which are able to work without specifying a model of a process to be optimized.

#### 3.2.2.1 Stochastic Approximation

A wide class of such methods has their origins in the seminal papers of Robbins and Monroe (RM) [145] and Kiefer and Wolfowitz [68]. In both of them the following regression model is considered:

$$m(\vec{x}) = \int_{R^{d_y}} y\, f_y(y;\, \vec{x})\, dy = \mathbb{E}_{Y(\vec{x})}\,[Y(\vec{x});\, \vec{x}]. \tag{3.8}$$

In [145] the problem of finding the solution of $m(\vec{x}) = 0$ is considered, while in [68] the minimum of $m(\vec{x})$ is to be found. The difficulty is in that $m(\vec{x})$ is unknown and the only available information is: for selected $\vec{x}_k$'s one can only acquire

$$y_k \text{ drawn at random according to } f_y(y;\, \vec{x}_k) \tag{3.9}$$

from a process (an environment).

Notice that $y_k$'s can always be expressed as their expectation plus a random component, i.e.,

$$y_k = m(\vec{x}(k)) + \varepsilon(k), \quad k = 1,\, 2,\, \ldots, \tag{3.10}$$

where random variable $\varepsilon(k)$ has a zero mean value, but its variance and higher moments, if they exist, may depend on $\vec{x}(k)$.

In the simplest setting, the Kiefer-Wolfowitz (KW) learning algorithm has the following form:

$$\vec{x}(k+1) = \vec{x}(k) - \gamma_k\, \widetilde{\nabla}_{\vec{x}}\, J(\vec{x}(k)) \tag{3.11}$$

where $\widetilde{\nabla}_{\vec{x}}\, J(\vec{x}(k))$ is an approximation of $\text{grad}_{\vec{x}} J(\vec{x})\Big|_{\vec{x}=\vec{x}(k)}$ by the central differences, i.e., $\widetilde{\nabla}_{\vec{x}}\, J(\vec{x}(k))$ is given by

$$\frac{1}{2\, c_k} \begin{bmatrix} (J(\vec{x}(k) + c_k\, \vec{e}_1) + \varepsilon_1'(k)) - (J(\vec{x}(k) - c_k\, \vec{e}_1) + \varepsilon_1''(k)) \\ (J(\vec{x}(k) + c_k\, \vec{e}_2) + \varepsilon_2'(k)) - (J(\vec{x}(k) - c_k\, \vec{e}_2) + \varepsilon_2''(k)) \\ \ldots \\ (J(\vec{x}(k) + c_k\, \vec{e}_N) + \varepsilon_N'(k)) - (J(\vec{x}(k) - c_k\, \vec{e}_N) + \varepsilon_N''(k)) \end{bmatrix},$$

where $\vec{e}_i, i = 1,\, 2,\, \ldots,\, N$ is the standard base of $R^N$, $\varepsilon_i'(k)$'s and $\varepsilon_i''(k)$'s are random errors in collecting observations of $J$. They are assumed to be zero mean, finite variance r.v.'s that are mutually independent with respect to all of their indexes. Step sizes $c_k > 0$ are selected as a sequence convergent to zero, but in a way corresponding to the rate of decay of $\gamma_k$'s sequence to zero. Namely, classic conditions imposed on these sequences are the following:

$$\sum_{k=1}^{\infty} \gamma_k = \infty, \quad \sum_{k=1}^{\infty} \gamma_k\, c_k < \infty, \quad \sum_{k=1}^{\infty} \gamma_k^2\, c_k^{-2} < \infty, \tag{3.12}$$

where $= \infty$ and $< \infty$ are the shorthand notations for divergent and convergent series, respectively. Under these conditions, extended by smoothness assumptions concerning $J$, it is possible to prove the convergence, as $k \to \infty$, of $\vec{x}(k)$ to $\vec{x}^*$, which is a point at which the gradient of $J$ is zero. The convergence is in the mean square sense and/or with the probability one, depending on assumptions.

RM and KW algorithms were pioneering stochastic learning algorithms. About 70 years have passed since their publication during which they were a subject of intensive theoretical studies and attempts of their applications. It is almost impossible to mention all the monographs and hundreds of papers on related topics. Therefore, only selected sources are referred to.

The following important issues were extensively studied.

**Convergence and convergence rates** The main drawback of the original KW algorithm was its slow convergence. Many attempts were undertaken to modify it so as to improve its asymptotic and practical speed ([41, 74, 102]).

There were also efforts directed toward non-asymptotic analysis [97] and to weakening convexity assumptions [7] as well as to obtain a global optimum seeking algorithm [37, 102] instead of finding stationary points.

**Incorporating constraints** The original KW algorithm was designed for unconstrained optimization problems. In many papers the orthogonal gradient projection is considered, while in [77] a counterpart of the feasible direction method is proposed.

**Applications in artificial intelligence algorithms** Modified versions of the RM and the KW algorithms are frequently used as building blocks of other learning algorithms. In particular, the RM approach is used in reinforcement learning methods, namely in Q-learning algorithms, in order to avoid estimating explicitly transition probabilities of the underlying Markov chain. In this context and for other issues of the RM and the KW algorithms the reader is referred to monographs [82, 83] and survey papers [28, 81].

Stochastic approximation plays also an important role in a subdiscipline called simulation optimization (see [46] for a comprehensive survey of this area of research), which is related to this book when model-based search methods for an optimal sequence of decisions are considered.

From the point of view of learning optimal decision sequences, the main drawback of the original KW algorithm is the necessity of running a process $2\,N$ times to perform one learning step. This can be too costly for many processes. As a remedy, one can use one-sided differences instead of central ones for estimating $\widetilde{\nabla}_{\vec{x}}\, J(\vec{x}(k))$, reducing the number of trials to $N+1$, but for large $N$ even this reduction of the process runs may not be sufficient.

It seems that when one has a decision sequence, $\vec{x}^{\diamond}$ say, which is acceptable for running the process, and one wants to improve its quality by learning, then an appropriate version of a stochastic approximation algorithm can be its asynchronous version proposed in [19], see also [128]. Its essence is in estimating the elements of $\widetilde{\nabla}_{\vec{x}}\, J(\vec{x}(k))$ one by one, i.e., applying repeatedly decision sequences $\vec{x}^{\diamond} + c_k\, \vec{e}_i$ to the process. In [19] one can find sufficient conditions for convergence of this version of learning.

### 3.2.3 Remarks on Model-Free Learning of Decision Sequences

In recent years the term *model-free learning* has appeared more and more frequently in different contexts and having different meanings. It seems that completely model-free learning of algorithms is not possible. At least variables involved have to be named and available as observations of a process (an environment). Additionally, one may expect (or know) that there are some relationships between these variables, even if these relationships are not expressed in terms of mathematical formulas or statistical dependencies.

Furthermore, a certain goal of learning is frequently specified, being a part of a model in a broad sense. The goal may not be defined explicitly, as in the field of reinforcement learning, but then a teacher (trainer, critic, oracle) influences a learning process according to certain general rules.

The classic area of classifying patterns (pattern recognition) or images is based on *learning with a teacher or expert* who knows[2] how to classify a given object (behavior) to one of a finite number of classes. Also in this area the term *model-free learning* or—more frequently—the term *distribution-free learning* is widely used. It means that the underlying probability distributions of patterns from each class exist, but they are not known (see also Sect. 3.3). Notice, however, that assuming the existence of probability distributions, according to which patterns (images) are drawn, is a strong assumption. Indeed, it means that there is an underlying, time invariant phenomenon (mechanism—a natural or established process) which generates patterns or images.

It is worth distinguishing so-called one-class classification problems, also connected to novelty detection tasks (see e.g. [79, 120, 165]). Their specific feature is that only positive examples, members of this one class, are known and used for learning. Here again the terms *model-free learning* and *distribution-free learning* are frequently used.

The one-class pattern recognition approach provides a link between learning a classifier and learning the optimal (according to a specified goal function) decision sequences. Namely, optimal or close to optimal decision sequences may serve as positive examples for one-class classifier. One can also consider an opposite point of view, applicable for searching for a global minimum. Indeed, when an evolutionary algorithm produces very similar decision sequences, then one can suspect that the algorithm was trapped into a local optimum. In such a case detection of *novelty*, i.e., a sequence not similar to that previously generated can be an indicator of leaving out the trap.

In the area of clustering (grouping) patterns or images there is a large number of approaches under the common name *learning without a teacher*. However, methods and algorithms from this stream of research are based on defining a certain measure of a similarity (or a dissimilarity) between items (behaviors) to be grouped together

[2] Usually it is tacitly assumed that classifications made by a teacher are correct. However, as proved in [52], when a teacher is allowed to make errors at random it is still possible to construct a nonparametric learning algorithm that converges to the Bayes risk.

or clustered. Such a measure can also be interpreted as a model that, indirectly, plays the role of a teacher.

A large class of approaches appearing under the name *model-free* is in fact based on a model, but this model is selected from a largely over-parametrized class of models such as neural networks or neuro-fuzzy systems (see e.g. [149, 184]). Such models are called black-box[3] models. The essence of such approaches can be summarized as follows:

1. select a class of models that is sufficiently rich to be able to generate behaviors which are observed when a real process (environment) is running
2. collect observations of the real process, trying to invoke (or only observe) a sufficiently broad class of its behaviors,
3. select and run a learning procedure of tuning parameters of a neural network or other black box model,
4. simplify the obtained model by applying either hypothesis testing about parameters or use a penalty for over-parametrization such as the Akaike information criterion (AIC), the Rissanen minimum description length (MDL) criterion or their modifications (see, e.g., [2, 8, 53]).
5. use the obtained model for learning decisions.

Variants of these approaches allow for updating parameter estimates also during the decision-making phase. In particular, the approach comprehensively described in [58] is based on intensive re-estimation of time-varying parameters that can be interpreted as sensitivity coefficients in a local linearization, which are updated at every time step.

Summarizing, at least a certain conceptualization seems to be necessary in order to be able to think about learning algorithms. Furthermore, learning decision sequences is even more demanding, since one has to preserve ordering of decisions in time. For these reasons it is worth discussing in more detail the possible roles of models in learning.

### *3.2.4 Possible Roles of Models in Learning*

In the previous sections the roles of model-free, model-based and model-inspired approaches for learning were already mentioned. Here, these topics and other possibilities are discussed in more detail.

**Model-based approach** In model-based approaches to learning decision sequences the main source of information for learning is a model itself. One

[3] The name *black-box* models was coined as opposite to models that are based on laws of the physics and chemistry. The term *gray box* models is also in use to designate models that are partly based on laws of physics and chemistry and partly on the estimation of unknown model ingredients, e.g. constitutive laws. In the context of nonparametric estimation such approaches are called semi-parametric models.

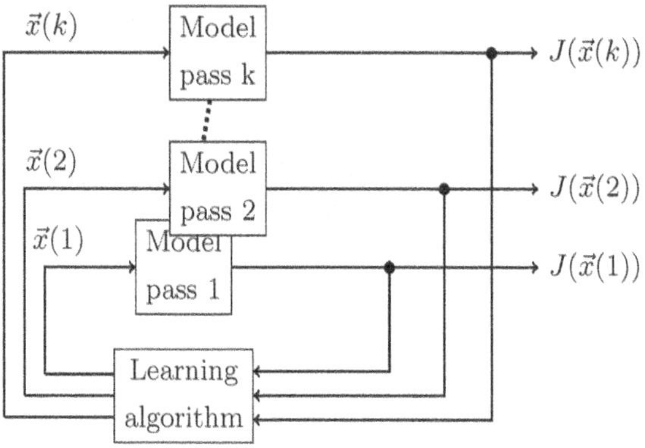


**Fig. 3.2** An outline of a repetitive learning algorithm interacting with a model

may (or must) heavily rely on the mathematical model at least in the following cases.

- When the model provides full information and simultaneously searching for optimal solutions is NP-hard. Examples were given in the previous sections (dynamic traveling salesman and operations scheduling problems).
- The laws of physics and chemistry provide a reliable model and experiments on a real process are very expensive and it is reasonable to replace them by simulation experiments (see [28]).
- The available model is simplified, but it is the only admissible source of data for learning. A case study on mitigating the spread of COVID-19 is provided in Chap. 5 as an illustration of this class of learning tasks. Inadmissibility of experimenting with this process is self-explanatory. For similar reasons models are used in simulating processes concerning nations' economies.

An outline of this kind of learning is shown in Fig. 3.2.

**Learning supported by a model** [16, 60]
An outline of model-supported learning is shown in Fig. 3.3.

**Model inspired learning**
**Learning from data driven models**

- concepts,
- functions,
- decision sequences,
- policies—decision rules

## 3.3 Plug-In Versus the Nonparametric Approach to Learning

A frequent way of designing decision-making algorithms is as follows:

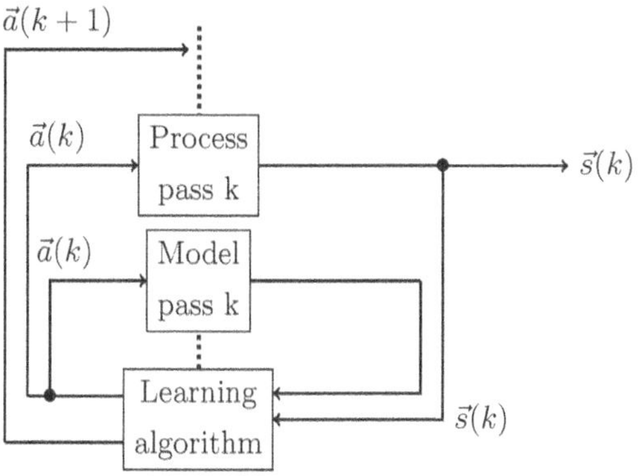


**Fig. 3.3** An outline of a repetitive learning algorithm interacting with a process and with a model

1. to assume that all ingredients necessary for deriving an algorithm e.g. a model, a goal function, probabilistic description of an environment are known,
2. to derive the optimal, or at least satisfactory, algorithm for solving a given problem,
3. to perceive that in practice some of the assumptions made in (1) are violated and to design a learning algorithm which aims at replacing unknown elements by their empirical counterparts that are based on observations of a process (an environment) and/or on historical data.

In the simplest cases, the above procedure leads to obtaining (in step 2) the optimal algorithm that contains unknown parameters, which originally were present either in the model and/or in probability distributions of a random environment. Then, learning these parameters is invoked—off-line or on-line—and the result is plugged-in into the optimal algorithm.

A competitive approach is known as the nonparametric one. Its essence is based on a direct approximation or estimation of a decision rule, without assuming any finite parametrization of its mathematical description. Approaches of this kind are also known as model-free.

Intermediate approaches are called semi-parametric. Roughly speaking, one is assuming a certain parametric model, but admittedly, this model may not be exact. As a consequence, a certain model extension is allowed and the extended part is specified in a nonparametric way, i.e., without assuming any finite parametrization of its functional form.

It should be added that nonparametric and semi-parametric approaches to learning are considered as asymptotic theories. In computational practice, they differ from the parametric approaches in that the number of learnable parameters is not specified in advance and is inferred from a learning sequence. Furthermore, the number of learnable parameters is allowed to grow when the number of observations is growing.

# Chapter 4
# Differential Evolution with a Population Filter

The differential evolution is a metaheuristic optimization method that allows the search for the global optimum. In comparison to the other metaheuristics, it requires a relatively smaller number of calculations. This can be helpful when a model is not available and experiments have to be carried out on a real system. In the decision-making context, this method can be used on relatively short decision sequences.

As a vehicle for presenting the idea of applying the population filter to handle constraints, the differential evolution algorithm is used in this chapter. At present, it is the algorithm of choice among all evolutionary type algorithms. However, it should be emphasised that any other evolutionary algorithm can be used instead.

On the other hand, one can consider the process of deriving new evolutionary algorithms, e.g. by "mutations" and combinations of known algorithms, as the upper-level evolutionary strategy. As a consequence, one must allow for applying not only the best algorithms as candidates for "mutations" and combinations in the next generation. From a psychological point of view, this opinion can be difficult to accept by researchers deeply involved in deriving better and better algorithms, but it is necessary if we want to keep the paradigms of evolutionary optimization also at the upper level.

## 4.1 Filter

The idea of a filter was proposed by Fletcher and Leyffer in [44] in order to avoid the direct usage of a merit function that modifies the goal function $f(\vec{x})$. This means that the problem of infeasible solution has to be solved in some other way. Let us define a function that measures how much the equality constraints $g_i(\vec{x})$ and inequality constraints $h_j(\vec{x})$ are violated.



W. Rafajłowicz, *Learning Decision Sequences For Repetitive Processes—Selected Algorithms*, Studies in Systems, Decision and Control 401,
https://doi.org/10.1007/978-3-030-88396-6_4

$$c(\vec{x}) = \sum_i |g_i(\vec{x})| + \sum_j \max(0, h_j(\vec{x})) \tag{4.1}$$

It is obvious that when $\vec{x} \in X$ so all constraints are met, then the newly defined $c(\vec{x}) = 0$. If the set $X$ is not empty, then there exists such $\vec{x}_f \in X$ that

$$c(\vec{x}_f) = 0. \tag{4.2}$$

We must also note that $\forall_{\vec{x}} c(\vec{x}) \geq 0$. This allows us to reformulate the problem (4.11)–(4.14) into the following one

$$\min_{\vec{x}} f(\vec{x}), \qquad \min_{\vec{x}} c(\vec{x}). \tag{4.3}$$

Obviously the problem (4.3) is a multi-objective (or multicriterial) optimization problem. The typical solving method calls for assigning weights[1] to each objective function.

$$\min_{\vec{x}} \left(f(\vec{x}) + \gamma c(\vec{x})\right). \tag{4.4}$$

After substituting (4.1) into (4.4) we obtain penalty function similar to simple, exterior penalty[2]:

$$\min_{\vec{x}} \left( f(\vec{x}) + \gamma(\sum_i |g_i(\vec{x})| + \sum_j \max(0, h_j(\vec{x}))) \right). \tag{4.5}$$

Following the idea of Fletcher and Leyffer [44], let us define the term *dominance* in the context of two-objective optimization. For any $\vec{x}_k$ we have pair $(f_k = f(\vec{x}_k), c_k = c(\vec{x}_k))$. A pair $(f_l, c_l)$ is said to be dominated by $(f_k, c_k)$ if and only if

$$c_l \geq c_k \text{ and } f_l \geq f_k \tag{4.6}$$

and at least one inequality has to be strict.

The filter is a data structure holding such pairs that they do not dominate one another. The following rules are kept in force when the content of filter $F_k$ is formulated at $k$th iteration of searching.

1. Filter $F_k$ contains only such pairs $(f_l, c_l)$ that do not dominate one another.
2. A new pair $(f_l, c_l)$ can be added to the filter if it is not dominated by any other pair already in the filter.[3]
3. Items, already in the filter, that are dominated by new $(f_l, c_l)$ will be removed from the filter in order to keep it agreeing with 1.

---

[1] In a two-criterial problem only one.

[2] The use of an absolute value instead of a square gives a nondifferentiable problem.

[3] Requirement of at least one strict inequality guarantees that every entry is distinct.

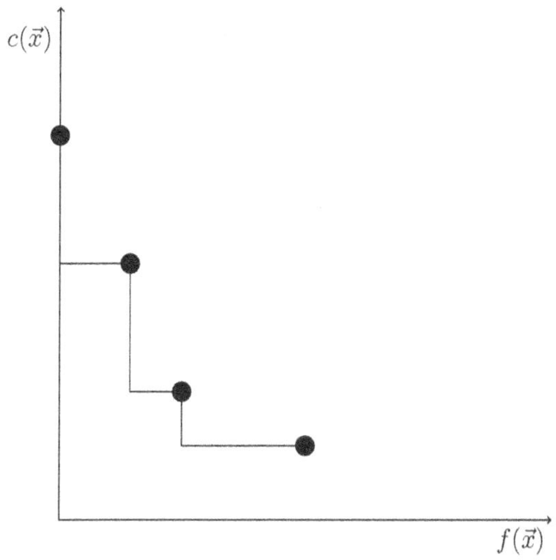


**Fig. 4.1** Pareto-like front forming in the filter. The circles represents entries in the filter

When rules 1–3 are satisfied the filter contents look as shown in Fig. 4.1. We must note here that the optimal solution of the constrained optimization problem is the pair $(f(\vec{x}^*), 0)$. This entry should not be dominated by any other entry in the filter.

The region to the right of horizontal-vertical lines in Fig. 4.1 is where values cannot be accepted into the filter.

The role of this data structure is to decide whether the result of a new iteration of optimization method should be accepted or discarded. A point that is accepted is that it has to be better in one way or another—better minimize goal function $f(\vec{x})$ or less violate the constraints.

## 4.2 Filter in Global Optimization Problems

In the previous chapter the idea of the Fletcher filter for handling constraints in smooth optimization problems was presented together with its further modifications.

### *4.2.1 Evolutionary Computations*

#### 4.2.1.1 Introduction

Evolutionary algorithms is the well-known class of global optimization methods. This group of methods tries to mimic and model evolution in living organisms.

The simplest, yet most effective part of this class of methods are phenotypical evolution methods [47, 48], see [49, 66] for recent monographs and [104] for interesting application.

Survey of constraints handling was presented in [92] and, apart from the method presented in subsequent sections, is still valid.

In subsequent sections the $i$th element of $k$th generation would be denoted by

$$\vec{x}_i(k). \tag{4.7}$$

This vector consist of scalar elements. In some algorithms it is necessary to manipulate each of those dimensions. Then the $l$th dimension would be denoted by

$$\vec{x}_{i(l)}(k). \tag{4.8}$$

#### 4.2.1.2 Traditional Algorithm

In the context of evolutionary computing the goal function is known as the fitness function and the problem is to maximize it. The typical formulation has to be changed in a simple way.

$$\min f(\vec{x}) = -\max[-f(\vec{x})] \tag{4.9}$$

Another requirement is that the goal function is positive.

Phenotype evolution treats $\vec{x}$ as a phenotype and the goal function as the fitness function. The typical framework for the method is as follows:

**Step 0** Get initial population $P(0) = \{\vec{x}_1, \dots \vec{x}_n\}$ where $n$ is the population size.
**Step 1** Calculate fitness function of all individuals.
**Step 2** Until the new population $P(k+1)$ is smaller than $n$,

1. *select* individual from population $P(k)$,
2. perform a mutation operation,
3. add the mutated individual to the new population $P^{(k+1)}$

**Step 3** Until STOP is reached go to Step 1.

The simplest method of selection is the so-called roulette-wheel selection, also known as proportional selection. Its goal is to make the probability of getting into next the generation proportional to individual fitness. To achieve this, the population is:

1. sorted according to fitness, giving new $f(\vec{x}_0) \geq f(\vec{x}_1) \geq f(\vec{x}_2) \geq \dots \geq f(\vec{x}_n)$ see Fig. 4.3,
2. adding in the following way $f(\vec{x}_0) + f(\vec{x}_1) + f(\vec{x}_2) + \dots + f(\vec{x}_{n-1}) + f(\vec{x}_n) \geq f(\vec{x}_1) + f(\vec{x}_2) + \dots + f(\vec{x}_{n-1}) + f(\vec{x}_n) \geq f(\vec{x}_2) + \dots + f(\vec{x}_{n-1}) + f(\vec{x}_n) \geq \dots \geq f(\vec{x}_{n-1}) + f(\vec{x}_n) \geq f(\vec{x}_n)$
3. normalized to [0, 1] by dividing with highest value $\sum_{i=0}^{n} f(\vec{x}_i)$,

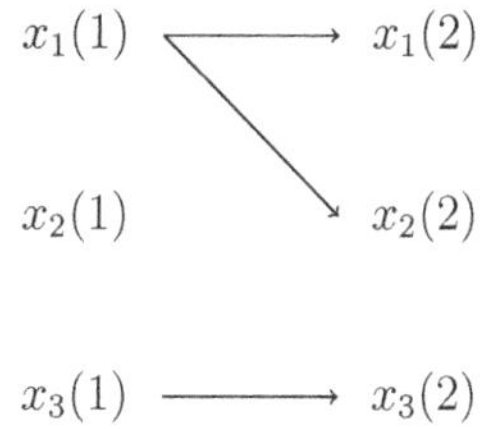


**Fig. 4.2** Example of a small population of three individuals in subsequent generations

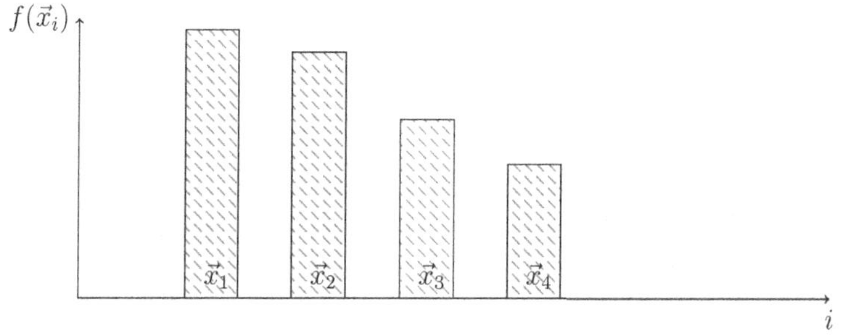


**Fig. 4.3** Population and its fitness

4. an individual is selected by choosing a random number [0, 1] and searching similar to an unequal sized wheel[4] see Fig. 4.4.

There exist many more methods, such as tournament selection, in Sect. 4.2.1.3 a modification with a constraints handling filter will be shown.

The mutation of a vector of numbers is also simple. Consider individual $\vec{x}_i(k)$

$$\vec{x}_i(k+1) = \vec{x}_i(k) + \vec{R}, \tag{4.10}$$

where $\vec{R}$ is a random vector from some, symmetric distribution—typically Gaussian distribution is used. The $\mu_i = x_i$ and $\sigma$ is a parameter.

The process is repeated until the desired size of the next generation is obtained. As a result we can see that some individuals have a different number of offspring in a new generation see Fig. 4.2.

This process results in the preference for better individuals in each generation. As a result in each subsequent generation a better result is achieved. Moreover, other membes of the population, possibly with worse fitness, can influence the next generation and find the global maximum.

[4] Since numbers are sorted, a binary search algorithm can be used.

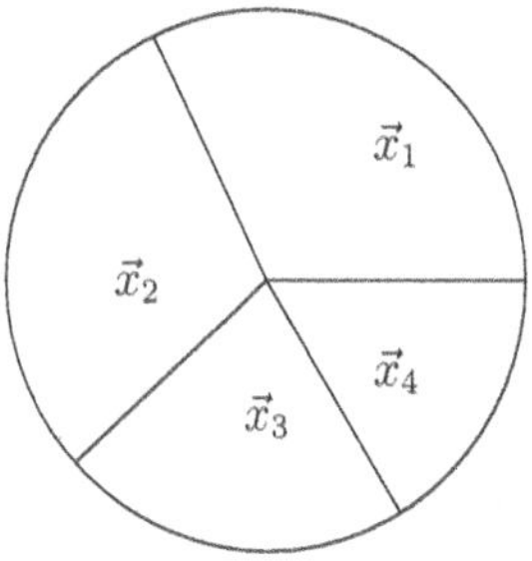


**Fig. 4.4** Normalized population for a proportional also known as a roulette-wheel selection

#### 4.2.1.3 Evolutionary Algorithms with the Filter for Handling Constraints

The methodology and general idea behind the filter is much wider than its use in sequential quadratic programming or sequential linear programming. It can be used in many other optimization methods as a method of handling constraints. In evolutionary methods the filter can be used as a soft selector which takes into account not only fitness of the individual but also the degree of constraints violation.

In *selection* the filter decides if the new point, generated by an evolutionary algorithm, should be accepted into the filter and also into new generation. This mechanism is *soft selection* because filter rules allow the new point to be accepted when the value of the objective function is worse than the best already found. This may happen only if the penalty for constraints violation decreases.

At the stop point the filter contains not only the best solution, but also the trade-off between objective function and constraints violation. In many cases we can allow for a small violation.

The selection mechanism can be softened even more if the filter is slightly modified. We can permit a dominated entry to stay in the filter for certain number of generations.

The problem solved has the following form:

$$\max f(\vec{x}) \quad \text{w.r.t.}\ \vec{x} \tag{4.11}$$

subject to constraints

$$f(\vec{x}) \geq 0, \tag{4.12}$$

$$g_i(\vec{x}) = 0, \tag{4.13}$$

$$h_j(\vec{x}) \leq 0 \tag{4.14}$$

**Evolutionary algorithm with the filter**

**Preparations** Select $n_{max}$—the largest possible population size and $n_{min}$—the smallest population that is reasonable. Choose $c_{max}$ and $f_{ub}$ in order to prevent unnecessarily large entries.

**Step 0** Set the counter for generations $k = 0$.

- Allocate an empty filter (list) $\bar{F}(0)$ and enter the triples $(dummy,\ -\infty,\ h_{max})$ and $(dummy,\ f_{ub},\ 0)$ to $\bar{F}(0)$, where $dummy$ stands for an arbitrary $d$-dimensional vector, which is not later used, but it must be added to maintain the compatibility of the filter entries.
- If possible, select $\vec{x}_0 \in X$, set $h_0 = 0$, $f_0 = f(\vec{x}_0)$. If $f_0 < f_{ub}$, enter $(\vec{x}_0,\ f_0,\ 0)$ into $\bar{F}(0)$, set $k = 1$ and go to Step 2. Otherwise, repeat the selection of $\vec{x}_0 \in X$.
- If it is not possible to select $\vec{x}_0 \in X$, go to Step 1.

**Step 1** **Initial population.** Choose $\gamma > 0$ as a tolerance of the constraints violation.

1. Select at random an initial population of points, admitting non feasible points, i.e., those outside $C$.
2. Run "a standard evolutionary algorithm" that does not take constraints into account, but minimizes $h(\vec{x})$ only as the goal function.
3. Stop this algorithm, if at least one point, $\vec{x}_0$ say, with $h(\vec{x}_0) \le \gamma$ is found.
4. Set $h_0 = h(\vec{x}_0)$, $f_0 = f(\vec{x}_0)$ (or $h_0'$, $f_0'$, $h_0''$, $f_0''$ etc., if there is more than one point with the penalty not exceeding $\gamma$).
5. Enter $(\vec{x}_0,\ f_0,\ h_0)$ into $\bar{F}(0)$ (or $h_0'$, $f_0'$, $h_0''$, $f_0''$ etc., according to rules regarding filter (see Sect. 4.1). Set $k = 1$.

**Step 2** **Trimming the filter.** Denote by $card[F(k)]$ the size[5] of filter $F(k)$, which is simultaneously the size of a current population.

1. If $n_{min} \le card[F(k)] \le n_{max}$, go to Step 3
2. If $card[e\ F(k)] < n_{min}$, apply random mutations of a moderate size to $\vec{x}_k', \vec{x}_k'', \ldots$ that are already in $F(k)$. Denote the results of mutations as $mut(\vec{x}_k')$, $mut(\vec{x}_k'')$, ...
3. Calculate $(f(mut(\vec{x}_k')),\ h(mut(\vec{x}_k'))), (f(mut(\vec{x}_k'')),\ h(mut(\vec{x}_k''))), \ldots$ and confront them with the current filters' content, according to rules regarding the filter. Repeat these steps unless $card[F(k)] \ge n_{min}$, then go to Step 3.
4. If $card[F(k)] > n_{max}$, sort the entries of $F(k)$ according to increasing penalties $h_k$'s and leave in $F(k)$ only the first $n_{max}$ entries. Go to Step 3.

**Step 3** **The next generation.** To each $\vec{x}_k', \vec{x}_k'', \ldots$ that are already in $F_k$ apply the following operations:

**offspring:** replicate each individual $\vec{x}_k', \vec{x}_k'', \ldots$ to $n + Round[n/(1 + h_k')], n + Round[n/(1 + h_k'')], \ldots$ number of descendants, respectively,

**mutations:** to each replica add a random vector and calculate the corresponding value of $f$ and $h$.

[5] Cardinality of a set.

**Table 4.1** Best values of the goal function in subsequent epochs, but only those for which $h(\vec{x}_k) \leq 0.005$ (left panel). Basic statistics from 30 simulation runs from $f$ values, but only those with $h < 0.005$ (right panel)

| Epochs | 1000 | 3000 |
|---|---|---|
| Min $f$ | 7161.1 | 7161.2 |
| Max $f$ | 7385.9 | 7266.1 |
| Median $f$ | 7264.0 | 7226.0 |
| Mean $f$ | 7265.5 | 7224.7 |
| Dispersion $f$ | 50.0 | 31.0 |

**soft selection:** Confront the results $(f, h)$ of the replications and mutations with the current filter contents and enter (reject) those, which are (not) in agreement with filter rules.

**Check the stopping condition.** If $k$ does not exceeds the admissible number of generations, set $k = k + 1$ and go to Step 2.

Simulation results were conducted on a typical benchmark for constrained global optimization from a set published by Schittkowski [157]. The tests were conducted for G08 problem

$$f(\vec{x}) = \frac{\sin^3(2\pi x_{(1)}) \sin(2\pi x_{(2)})}{x_{(1)}^3 (x_{(1)} + x_{(2)})}, \tag{4.15}$$

subject to

$$g_1(\vec{x}) = x_{(1)}^2 - x_{(2)} + 1 \leq 0, \tag{4.16}$$

$$g_2(\vec{x}) = 1 - x_{(1)} + (x_{(2)} - 4)^2 \leq 0. \tag{4.17}$$

The results are summarized in Table 4.1.

## 4.2.2 *Differential Evolution*

Differential evolution is a metaheuristic method of solving the global optimization problem. The method was originally proposed by Storn and Prince in 1995. The name "evolution" would suggest some nature-inspired algorithm. In this case there was no direct inspiration by some process but the general outline of the method is similar.

Differential evolution is an ongoing research topic. A large number of papers as well as monographs were published, among the latter [27, 42, 126].

#### 4.2.2.1 Original Method

The method is based on a population consisting of $n$ vectors[6] $\vec{x}_0(k), \ldots, \vec{x}_n(k)$, $k$ is the generation/iteration of a method. In this section then notation $x_{(i)}$ is the $i$th element of vector $\vec{x}$.

The basic method depends on the following parameters

- $n$—size of the population,
- $d$—size of the $\vec{x}_i$ vector,
- $CR \in [0, 1]$—crossover probability,
- $F \in [0, 2]$—mutation scale factor, also called differential weight.

In each iteration a new population is generated in the following way for each member of population $\vec{x}$.

**selecting elements** A random element is selected from the population, let us call it $\vec{b}$. Next, two others are selected, such that $\vec{a} \neq \vec{b} \neq \vec{c}$. Select current working direction $R$—a random number $1, \ldots, m$.

**mutation** In every dimension $l = 1 \ldots d$:

1. select random number $r \in [0, 1]$, usually from the uniform distribution,
2. if $r < CR$ or when in the working dimension ($l = R$) then $y_{(l)} = a_{(l)} + F \cdot (b_{(l)} - c_{(l)})$
3. otherwise $y_{(l)} = x_{(l)}$.

**selection** if $f(\vec{y}) < f(\vec{x})$ then $\vec{x} = \vec{y}$, check new $\vec{x}$ for being a new, best solution.

As can be noticed, in each iteration and for every vector, a check is performed in order to find the global best result. The large number of starting points (in the initial population) and random selection of the $\vec{a}, \vec{b}, \vec{c}$ points allows for exploration of the solution space.

Simplicity of implementation and ease of parallelization have made differential evolution a popular algorithm.

The proposed mutation is one of many versions presented in papers, see [3, 63, 108].

#### 4.2.2.2 Differential Mutation as a Random Gradient

The differential mutation $\vec{a}_{(l)} + F \cdot (\vec{b}_{(l)} - \vec{c}_{(l)})$ can, in specific circumstances, be seen as a modified approximation of the gradient. When the population is around the optimum—which is a typical situation near the end of successful optimization—the situation similar to that shown in Fig. 4.5 can be seen.

The difference $\vec{b} - \vec{c}$ gives a good direction to the optimum. When anchored at $\vec{a}$, the resulting point $\vec{a} + F \cdot (\vec{b} - \vec{c}$ is a very good candidate for optimum as seen in Fig. 4.6.

---

[6] At this point, not dissimilar to evolutionary computing.

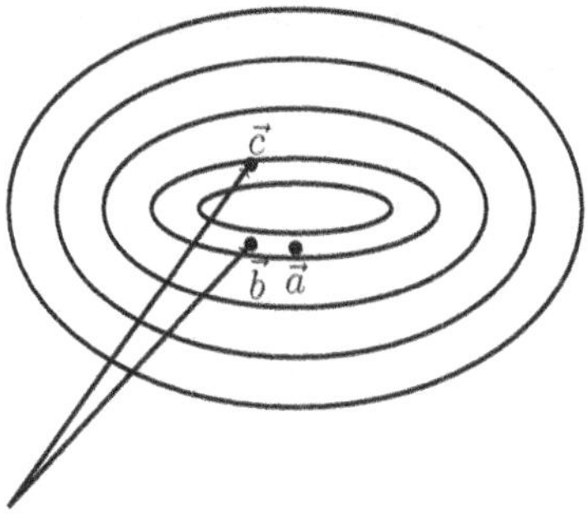


**Fig. 4.5** Three points around the optimum in differential evolution

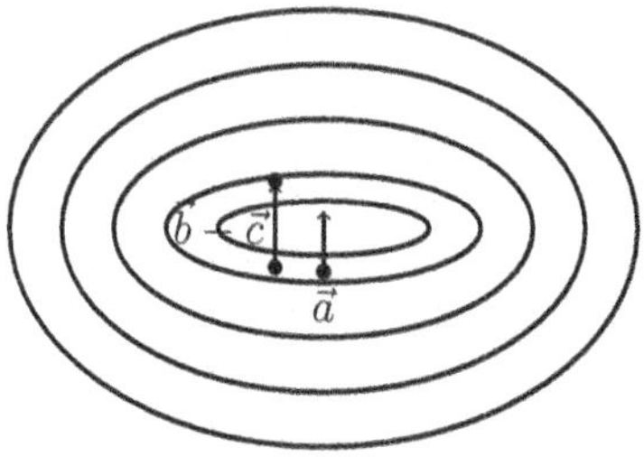


**Fig. 4.6** Difference between points and $\vec{a} + F(\vec{b} - \vec{c})$, $F = 0.5$

#### 4.2.2.3 Applying the Filter

In the paper [134] the author has proposed a filter-based, constraints handling method. Many other methods were proposed, but most of them are only capable of box-type constraints, see [11] or [5]. In the FilterDE method a population consists of elements in the filter. This means that the population size changes in each iteration—usually growing. Usually the solution can be found in the filter as the one having the smallest or zero constraints violation $c(\vec{x})$.

This leads to the following algorithm:

**Differential evolution with filter**

1. Select a suitable population/filter—the simplest method is to use a reasonable metaheuristic method, even differential evolution itself, to minimize function $c(\vec{x})$, the resulting *final population*[7] is added to the filter—dominated solutions are removed.
2. Choose parameters $F$, $CR$ as in the non-modified method presented in Sect. 4.2.2.1.
3. Until some STOP criterion is satisfied, for each $\vec{x}$ in the population/filter:

**Step 1** do *mutation* in the way described previously, obtain a candidate $\vec{y}$.
**Step 2** *selection*—for the resulting $\vec{y}$ calculate the pair of criterion value and infeasibility $(f(\vec{y}), c(\vec{y}))$, if this pair is accepted into the filter, replace $\vec{x} = \vec{y}$, otherwise do not change the population.

[7] Not only the best result.

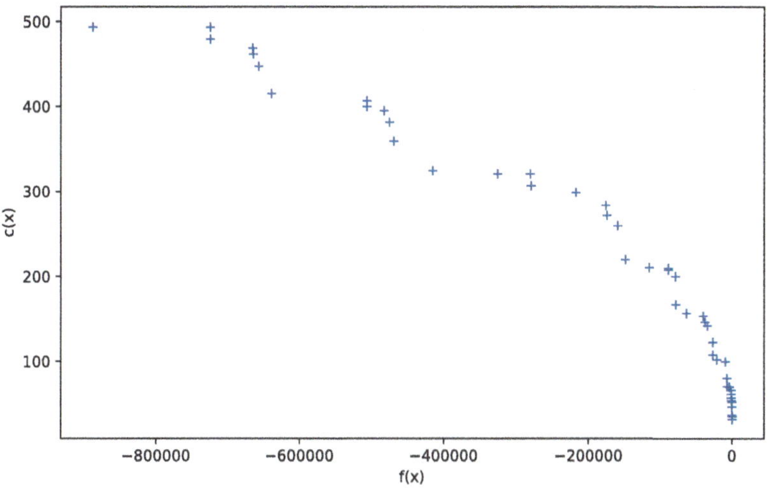


**Fig. 4.7** The resulting, final filter content for the G01 problem

The method was tested on a typical benchmark example `G01` proposed in [157]. This is a minimization problem with 13 variables.

$$f(\vec{x}) = 5\sum_{i=1}^{4} x_{(i)} - 5\sum_{i=1}^{4} x_{(i)}^2 - \sum_{i=5}^{13} x_{(i)} \tag{4.18}$$

subject to

$$g_1(\vec{x}) = 2x_{(1)} + 2x_{(2)} + x_{(10)} + x_{(11)} - 10 \leq 0$$

$$g_2(\vec{x}) = 2x_{(1)} + 2x_{(3)} + x_{(10)} + x_{(12)} - 10 \leq 0$$

$$g_3(\vec{x}) = 2x_{(2)} + 2x_{(3)} + x_{(11)} + x_{(12)} - 10 \leq 0$$

$$g_4(\vec{x}) = -8x_{(1)} + x_{(10)} \leq 0$$

$$g_5(\vec{x}) = -8x_{(2)} + x_{(11)} \leq 0$$

$$g_6(\vec{x}) = -8x_{(3)} + x_{(12)} \leq 0$$

$$g_7(\vec{x}) = -2x_{(4)} - x_{(5)} + x_{(10)} \leq 0$$

$$g_8(\vec{x}) = -2x_{(6)} - x_{(7)} + x_{(11)} \leq 0$$

$$g_9(\vec{x}) = -2x_{(8)} - x_{(9)} + x_{(12)} \leq 0$$

additional bounds are $0 \leq x_{(i)} \leq 1$ for $i = 1, \ldots, 9, 13$ and $0 \leq x_{(i)} \leq 100$ for $i = 10, 11, 12$.

In Fig. 4.7 the final filter after 1000 iterations is shown.[8]

A modification can be proposed. In step 2, selection can occur in many ways. The simplest one is to accept a new solution, that is $x = y$, only when $y$ dominates $x$. The filter is used only to keep the best, in a Pareto-optimal context

[8] Please note that some points are very similar in $c(x)$ value and appear to be one above the other. This is a result of the plot and they are not dominated.

# Chapter 5
# Decision Making for COVID-19 Suppression

The main aim of this chapter is to discuss the possibilities of rational decision making for mitigating the spread of COVID-19. It also illustrates the basic notions introduced in Chap. 2 and the advantages of the modified differential evolution with the population filter proposed in Chap. 4.

## 5.1 Modified Logistic Growth Model and Its Validation for Poland

As a vehicle for the presentation a simple model of a pandemic, the spread of SARS Cov-2 is discussed. Due to its simplicity, it is also possible to formulate the problem of learning decision sequences that minimize the number of infected people, taking into account the costs of decisions and constraints. It seems worth considering, since the main stream of research concentrates on the prediction of the spread of COVID-19 (see [15] and the bibliography cited therein). The learning algorithm and the results of its testing are provided in the end of this chapter.

Before going into detail, one can ask the question whether the COVID-19 pandemic is a repetitive process. Unfortunately, the answer is positive for the following reasons:

- very similar patterns of the spread of SARS CoV-2 can be observed in different countries or states (provinces) of large countries
- it was earlier expected and now confirmed that COVID-19 outbreaks reappear in countries where its end was already expected (see Fig. 5.1).



W. Rafajłowicz, *Learning Decision Sequences For Repetitive Processes—Selected Algorithms*, Studies in Systems, Decision and Control 401,
https://doi.org/10.1007/978-3-030-88396-6_5

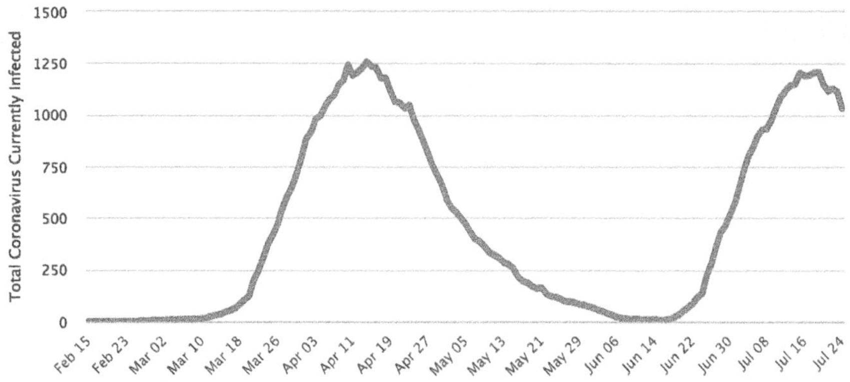


**Fig. 5.1** Daily numbers of infected in Croatia as a repetitive process

### *5.1.1 Classic Logistic Growth Models*

The logistic growth models seem to be the simplest that are able to catch the basic features of the phenomenological behavior of epidemics, including the spread of COVID-19 (see [123, 173, 183] and the bibliography cited therein).

Denote by $s(t)$ the number of infected at time $t$ in a certain country. The classic logistic model is given by the following differential equation:

$$\frac{d\,s(t)}{dt} = r\,s(t)\left(1 - \frac{s(t)}{K}\right), \quad s(0) = s_{init}, \tag{5.1}$$

where $K > 0$ is called an environment capacity, while $r > 0$ is the epidemic growth rate, which is measured in $1/timeunit$. The initial condition $s_{init}$ is the number of infected when simulations are starting.

More flexibility in reproducing the epidemic growth and decay phases is provided by the generalized Richards model:

$$\frac{d\,s(t)}{dt} = r\,s^{p}(t)\left[1 - \left(\frac{s(t)}{K}\right)^{\alpha}\right], \quad s(0) = s_{init}, \tag{5.2}$$

where $0 \leq p \leq 1$ and $\alpha > 0$ are parameters that can be tuned in order to obtain a good fit to observations. Observe that in this model, the parameter $r$ is measured in the unit of $1/day^{p}$, to keep the dimensional integrity of the left and the right-hand side of (5.2).

### 5.1.2 Modified Logistic Growth Model

The following model is proposed in this book:

$$\frac{d\,s(t)}{dt} = \frac{\hat{r}}{\tau}\,s(t)\left[1 - \frac{s(t)}{K}\right]^{\beta}, \quad s(0) = s_{init}, \tag{5.3}$$

where $\beta > 0$ is a tuneable parameter, $s_{init} < K$. Its value for the initial phase (94 days, starting from the 1st of April 2020) of the spread of COVID-19 in Poland was selected to be $\beta = 2$ in order to properly reflect a relatively slow growth of the number of infected at this phase. In (5.3) $\tau > 0$ is the average time that passes from when an individual is detected as an infected to recovery (assumed to be 21 days[1] in further simulations). In the simulations $K = 40000$ and $s_{init} = 2946$ persons were infected. The epidemic growth rate $\hat{r} > 0$ is a dimensionless parameter, tuned to $\hat{r} = 1.08$. Notice that $\hat{r}$ differs than the well-known reproduction rate $R$ parameter. Notice that the left and the right-hand side of (5.3) are dimensionally compatible.

### 5.1.3 Bernstein Polynomials as Possible Models of the Epidemic Growth Rate

The right-hand sides of (5.1) and (5.3) can be interpreted as special cases of Bernstein polynomials, after an appropriate scaling. Indeed, for $\mathcal{N}$ being the set of nonnegative integers, define $p$th Bernstein polynomial[2] of the order $q$, $p \le q$, $p$, $q \in \mathcal{N}$ as follows (see e.g. [88])

$$B_q^{(p)}(b) = \binom{q}{p} b^p\,(1-b)^{(q-p)}, \quad b \in [0,\,1], \quad p = 0,\,1,\,\ldots,\,q. \tag{5.4}$$

Setting $b = s(t)/K$ one can notice that the right hand sides of (5.1) and (5.3) are proportional to $B_2^{(1)}(s(t)/K)$ and $B_3^{(1)}(s(t)/K)$, respectively. These polynomials are plotted in Fig. 5.2, together with $B_4^{(1)}(b)$, in order to illustrate the flexibility of Bernstein polynomials as the epidemic growth rate models. As is known, the maximum of Bernstein polynomials appear at $b = p/q$, which allows for flexible modeling of the COVID-19 spread in different countries.

Consider the following model for the number of infected, when Bernstein polynomials are used as the epidemic growth rate:

[1] As mentioned in [119], "The median duration of viral shedding was 20 days in Wuhan inpatients".

[2] Symbols $p$ and $q$ are used locally in this section are unrelated to the same symbols that are used elsewhere in this book.

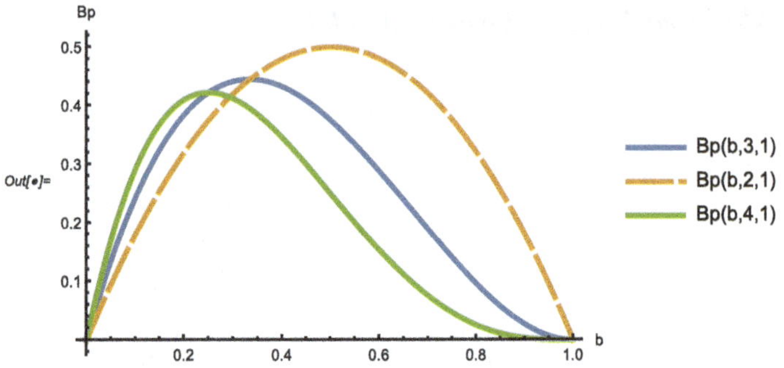


**Fig. 5.2** Selected Bernstein polynomials as possible models of the epidemic growth rate

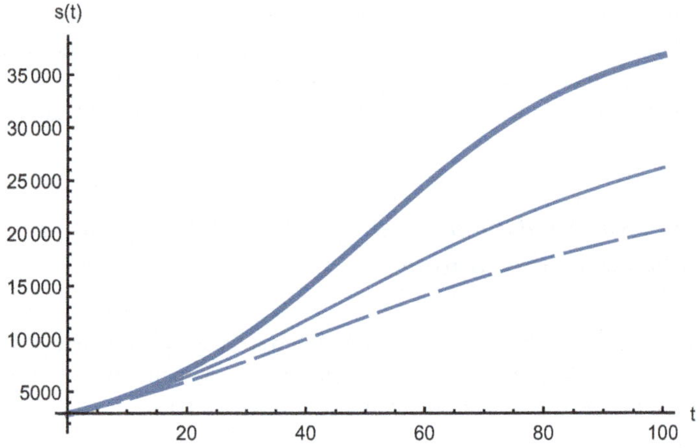


**Fig. 5.3** The growth of the number of infected when selected Bernstein polynomials (the same as in Fig. 5.2) are used as possible models of the growth rate

$$s'(t) = \hat{r}\,\tau^{-1}\binom{q}{p}^{-1} K^p\, B_q^{(p)}(s(t)/K)\,, \quad s(0) = s_{init}. \tag{5.5}$$

It leads to the growth of the number of infected as shown in Fig. 5.3.

For further considerations, model (5.3) with $\beta = 2$ (proportional to $B_3^{(1)}$) is selected.

### 5.1.4 *Model Discretization and Validation*

The next step in adapting (5.3) to the purposes of this book is its discretization over the equidistant grid $t_n$, $n = 0, 1, \ldots, N$ with step size $\Delta t > 0$. In general, a discretization of epidemic growth models requires advanced techniques (see [4] and

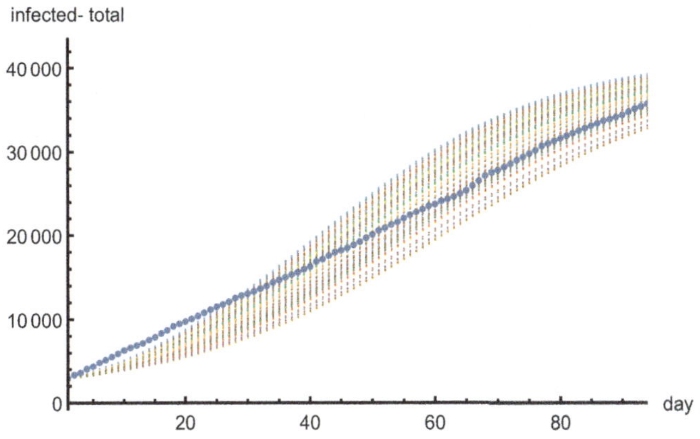


**Fig. 5.4** Numbers of infected in Poland, starting from 1st of April 2020 and 100 trajectories of the randomly perturbed model (5.6)

bibliography therein). Here, it suffices to use the simplest Euler scheme that leads to the following model:

$$s_{n+1} = s_n + \frac{\hat{r}}{\tau} s_n \left[1 - \frac{s_n}{K}\right]^\beta \Delta t, \quad s_0 = s_{init}, \quad n = 0, 1, \ldots, N, \tag{5.6}$$

where $s_n$ approximates $s(t_n)$. Further, $\Delta t$ is set to one day and it is not displayed.

In order to validate model (5.6) a sequence of $N = 94$ uniformly distributed in $[-50, 75]$ random variables[3] was subsequently added to the right-hand side of (5.6). These simulations were repeated 100 times. The results are plotted in Fig. 5.4 together with the observed numbers of infected in Poland in the same period. As one can observe, after about the first 20 days real-life data are contained in the confidence bands and one can use (5.6) for illustrative purposes.

## 5.2 Optimization of Decision Sequence—Problem Statement

Models of the COVID-19 pandemic usually consider parameter $\hat{r}$ as a constant. However, it seems to be justified to consider its values as the results of decisions made by authorities as well as everyday decisions taken by citizens.

[3] Random perturbations added to model (5.6) are not symmetric around zero, since one can expect that the announced number of infected persons is rather underestimated than overestimated.

### 5.2.1 Actions Reducing the Spread Of COVID-19

Although $\hat{r}$ is not equal the reproduction rate $R$, the qualitative behavior of both is similar, i.e., the reduction of $R$ reduces also $\hat{r}$ . A large number of actions that lead to the reduction of $R$ and $\hat{r}$ is is known, including:

- quarantine,
- maintaining social distancing,
- wearing face masks,
- intensive testing and isolation,
- lockdown

and others (see [119], where also the estimates of reducing $R$ are provided).

The above list of possibilities of reducing $R$ and $\hat{r}$ makes it rational that $\hat{r}$ can serve as a decision variable. Even more flexibility can be obtained if it is allowed to change $\hat{r}$ over time. Thus, later we replace $\hat{r}$ by $\vec{x}$=$[x_0, x_1, \ldots, x_N]^T$, which leads to the following model:

$$s_{n+1} = s_n + \frac{x_n}{\tau} s_n \left[1 - \frac{s_n}{K}\right]^\beta, \quad s_0 = s_{init}, \quad n = 0, 1, \ldots, N, \tag{5.7}$$

that is of the form (2.34) with

$$G(s_n, x_n) = s_n + \frac{x_n}{\tau} s_n \left[1 - \frac{s_n}{K}\right]^\beta. \tag{5.8}$$

### 5.2.2 Constraints

The following constraints are imposed on all $x_n$'s

$$x_{min} \le x_n \le x_{max}, \quad |x_{n+1} - x_n| \le \Delta x_{max}, \tag{5.9}$$

where $x_{min} \ge 0$ and $x_{min} < x_{max}$ are the lowest and the highest admissible levels of the decision variables, respectively, while $\Delta x_{max} > 0$ is the largest admissible change of decisions between subsequent days. In simulations of decision optimization problems presented later, $x_{min} = 0.5$, $x_{max} = 1.5$ and $\Delta x_{max} = 0.25$ were selected.

The rationale behind the first group of constraints is that forcing $x_n$ below 0.5 can be too expensive to a society, while allowing $x_n$ above 1.5 would lead to the dangerously fast growth of the epidemic, as has happened in many countries.

**Remark 5.6** (*Caution*) Additionally, it should be stressed that $x_n$'s near the upper bound (here 1.5) must be avoided unless the number of those infected falls to low levels e.g. below 10 in countries with larger populations.

The second group of constraints is aimed at preventing too hasty decision making, which can be difficult for societies to tolerate in the long term.

**Remark 5.7** (*Possible extensions*) One can consider also $K$ in (5.7) as a decision variable with the interpretation that the capacity of the medical system can be enlarged at some cost when necessary. In such a case actions $a_n$ would consist of $x_n$ and $K$ dependent on time, but later this extension is not considered.

### 5.2.3 *Interpreting the Goal Function*

The following expression is selected as a decisions quality criterion:

$$J(\vec{x}) = s_N + \lambda \sum_{n=0}^{N} \gamma^n (x_{max} - x_{n-1})^2, \tag{5.10}$$

$$\hat{J}(\vec{x}) = \frac{s_N}{K} + \hat{\lambda} \sum_{n=0}^{N} \gamma^n (x_{max} - x_{n-1})^2 = \frac{1}{K}\left(s_N + \hat{\lambda} K \sum_{n=0}^{N} \gamma^n (x_{max} - x_{n-1})^2\right), \tag{5.11}$$

where $\lambda = \hat{\lambda} K > 0$ and $\hat{\lambda}$ can be interpreted as the unit cost of reducing $x_n$. It is further selected as $\lambda = 50$, while $\gamma = 1.08$.

This goal function can be explained as follows. One aims to minimize the total number of infected persons at the end of the period $N$. Simultaneously, the costs $(x_{max} - x_n)^2$ of reducing $x_n$ from its upper admissible limit $x_{max}$ (hence, also reducing $R$) are taken into account.

**Remark 5.8** Notice that rising $(x_{max} - x_n)$ to the power two plays a different role than usual. Namely, $x_n$ is always less than or equal to $x_{max}$, but the squared value of $(x_{max} - x_n)$ lies below linear costs for $(x_{max} - x_n) < 1$.

The costs $(x_{max} - x_n)^2$ are discounted by the factor $\gamma^n$ with $\gamma > 1$ that aims to reflect possible impatience or even restlessness of the public due to unacceptably lengthy social restrictions.

Notice that (5.10) falls into the general scheme (2.40) by selecting

$$\phi_n(s_n, x_{n-1}) = \lambda \gamma^n (x_{max} - x_{n-1})^2, \quad n = 1, 2, \ldots, (N-1) \tag{5.12}$$

and $\phi_N(s_N, x_{N-1}) = s_N$.

Summarizing, the problem of selecting the optimal decision sequence $\vec{x}^*$ is of finding the global minimum of (5.10), under the constraints (5.7) and (5.9).

Notice that the process of the COVID-19 spread (5.7) is nonlinear with respect to the decision sequence. Furthermore, part of constraints (5.9) is not differentiable. Thus, one cannot hope to find an analytical solution to the above optimization problem. For the same reasons, many known methods of numerical search for the optimum can not be applied for finding its approximate solution.

## 5.3 Searching for Decisions Using Symbolic Calculations

Goal function (5.10) contains the term $s_N$, which is the prediction of the number of infected and detected people for $N$ days ahead as a function of decision sequence $\vec{x}$. One cannot expect that it will be possible to provide reliable predictions of $s_N$ from observed data only. Therefore, a model-based approach seems to be the only possibility of searching decisions. In this section a brief study of the possible use of symbolic calculations is presented.

### 5.3.1 *Model-Based Prediction by Symbolic Calculations*

The analysis of the model:

$$s_{n+1} = s_n + \frac{x_n}{\tau} s_n \left[1 - \frac{s_n}{K}\right]^2, \quad s_0 = s_{init}, \quad n = 0, 1, \ldots, (N-1), \tag{5.13}$$

reveals that $s_N$ is a polynomial in $(N-1)$ variables $x_0, x_1, \ldots, x_{N-1}$. The number of terms of such polynomials grows very rapidly, as shown in the first row in Table 5.1. In general, such a fast growth of the number of monomials in $x_0, x_1, \ldots, x_{N-1}$ precludes the use of purely a symbolic approach for the optimization with respect to these variables, even for a relatively small $N$, or makes it very time-consuming. The reason is that in purely symbolic computations $s_N$ is expressed both by variables $x_0, x_1, \ldots, x_{N-1}$ and by the model parameters $K, \tau$ and $s_{init}$. As an illustration consider $s_4$ with $s_{init}$ replaced by $s_0$ for the sake of brevity:

$$s_4 = s_0 \left( \frac{x_2(K - s_0)^2 \left(K^2\tau + x_1(K - s_0)^2\right) \left(K^2\tau - K x_1 s_0 + x_1 s_0{}^2\right)^2}{K^8 \tau^4} \right) +$$

$$+ s_0 \left( \frac{x_3 \left(x_2(K - s_0)^2 \, T1\right)}{K^{26}\tau^{13}} \right) + \tag{5.14}$$

**Table 5.1** The number of monomials of variables $x_0, x_1, \ldots, x_{N-1}$ for predicting $s_N$. Case 1—the total number of monomials when $s_N$ is computed using the symbolic method. Case 2—the number of monomials when $s_N$ is computed by the hybrid symbolic-numerical algorithm

| | $N$—pred. days | 2 | 3 | 4 | 5 | 6 | 7 |
|---|---|---|---|---|---|---|---|
| Case 1 | Total no terms | 2 | 6 | 34 | 438 | 13874 | 1102038 |
| Case 2 | Reduced no terms | 2 | 6 | 19 | 56 | 142 | 316 |

$$+ s_0 \left( \frac{\left(K^6 x_1 \tau^3 (K - s_0)^2 + K^8 \tau^4\right) T2}{K^{26} \tau^{13}} + \frac{x_1 (K - s_0)^2}{K^2 \tau} + 1 \right)$$

where

$$T1 \stackrel{def}{=} \left(K^2 \tau + x_1 (K - s_0)^2\right) \left(K^2 \tau - K x_1 s_0 + x_1 {s_0}^2\right)^2$$

$$T2 \stackrel{def}{=} \left(K^9 \tau^4 - s_0 \left(x_2 (K - s_0)^2 \left(K^2 \tau + x_1 (K - s_0)^2\right)\right)\right)^2 \times$$

$$\times \left(\left(\left(K^2 \tau - K x_1 s_0 + x_1 {s_0}^2\right)^2 + K^6 x_1 \tau^3 (K - s_0)^2 + K^8 \tau^4\right)\right)^2 .$$

The full form of $s_5$ would take two pages.

One possible way of circumventing these difficulties is to apply parallel computing (see [43] and the bibliography cited therein). An alternative method is proposed below.

Fortunately, one can indicate the class of problems for which symbolic calculations are not so time-consuming. Namely, these are tasks such that the coefficients of the high degree monomials in $x_0, x_1, \ldots, x_{N-1}$ are very small and, simultaneously, admissible ranges of changes of each $x_0, x_1, \ldots, x_{N-1}$ are bounded. In such cases one can apply a hybrid symbolic-numerical approach for predicting $s_N$ as a sum of monomials in $x_0, x_1, \ldots, x_{N-1}$. Its essence is summarized below.

**The hybrid symbolic-numerical algorithm for predicting $s_N$**

1. generate $s_N$ recursively as in (5.13), but
2. keeping $x_0, x_1, \ldots, x_{N-1}$ as symbolic variables and
3. simultaneously substituting numerical values of parameters (here: $K, \tau, s_{init}$),
4. performing numerical calculations on them in order to compute the coefficients corresponding to the monomials in $x_0, x_1, \ldots, x_{N-1}$ and
5. setting to zero the coefficients smaller than prescribed $\epsilon > 0$ and removing the corresponding monomials from further calculations.

Applying this procedure to (5.13) with $K = 40\,000$ and $\tau = 21$ essentially reduces the number of the terms needed for predicting $s_N$, as illustrated in the second row of Table 5.1. This reduction stems from the fact that the coefficients of the high degree terms contain the multipliers $K^{-p}$ with very high positive $p$'s. For example, polynomial $s_4$ contains a $K^{-26}$ multiplier before one of the monomials and formula (5.14) reduces to the following:

$$s_4 = -0.03\, x_2\, x_1^2 - 0.03 x_3\, x_1^2 + 4.13\, x_2\, x_1 + 0.08\, x_2\, x_3\, x_1 + \tag{5.15}$$

$$4.13\, x_3\, x_1 + 120.38\, x_1 + 120.38\, x_2 - 0.03\, x_2^2 x_3 + 4.13\, x_2\, x_3 + 120.38\, x_3 + 2946$$

when the coefficients are rounded to two digits after the decimal point and those less than 0.01 are set to zero.

The second important factor that differentiates purely symbolic calculations and hybrid ones is in the time of computations. The time needed for computing Case 1 row in Table 5.1, including the generation of $s_N$ polynomials, was about 3.2 hours, while for Case 2 only 10 seconds. In both cases computations were done using the Mathematica 12 environment run on a typical PC computer with 3.2 GHz processor and 32 GB of memory.

### *5.3.2 The Newton Method Using Hybrid Computations*

Recall the goal function (5.10)

$$J(\vec{x}) = s_N + \lambda \sum_{n=1}^{N} \gamma^n \left(x_{max} - x_{n-1}\right)^2 . \tag{5.16}$$

The gradient of the second summand is easy to compute in the symbolic way. For moderate $N$, the symbolic computations of the gradient of $s_N$ can be made in a reasonable time when the hybrid numerical-symbolic algorithm is previously applied.

Summarizing, for fixed parameters of the COVID-19 model, one can compute the gradient, denoted as $\text{grad}_{\vec{x}}\, J(\vec{x})$, as a sum of monomials composed from $\vec{x}$ elements, i.e.,

$$x_1^{\eta_1}\, x_2^{\eta_2} \ldots x_N^{\eta_N} \quad \eta_i \in \mathcal{N},\ i = 1, 2, \ldots, N. \tag{5.17}$$

Analogously, the partial derivatives of $\text{grad}_{\vec{x}}\, J(\vec{x})$ can be calculated in the symbolic way, leading to the Hessian matrix, further denoted as $H(\vec{x})$. As is known, $H(\vec{x})$ is an $N \times N$ symmetric and nonnegative definite matrix for every $\vec{x}$.

In order to store $J(\vec{x})$ in a functional form it suffices to store powers $\eta_i$'s and coefficients corresponding to each monomial. This remark applies to $\text{grad}_{\vec{x}}\, J(\vec{x})$ and $H(\vec{x})$ analogously.

Thus, one can use the well-known Newton method in the following form.

**The Newton algorithm in the hybrid numerical-symbolic form**

1. Choose $\epsilon > 0$ to be used in the stopping condition and $\gamma > 0$ as the updating step size. Select an initial sequence of decisions $\vec{x}(0)$ such that $H(\vec{x}(0))$ is positive definite and set pass (trial) number $k = 0$.
2. Run the hybrid symbolic-numerical algorithm for predicting $s_N$. Then, compute $J(\vec{x})$, $\text{grad}_{\vec{x}}\, J(\vec{x})$ as well as $H(\vec{x})$) and store them in the functional form.
3. Calculate the numerical version of the gradient of $J$ by substituting $\vec{x}(k)$ into $\text{grad}_{\vec{x}}\, J(\vec{x})$. Optionally, calculate also the numerical version of $J(\vec{x})$ at $k$th pass by substituting $\vec{x}(k)$. If

$$||\text{grad}_{\vec{x}}\, J(\vec{x}(k))|| \leq \epsilon, \tag{5.18}$$

then STOP, providing $\vec{x}(k)$ as the result. Otherwise, go to the next step.

4. Update $\vec{x}(k)$ as follows:
   (a) form the current Hessian matrix $H(k)$ by substituting $\vec{x}(k)$ into $H(\vec{x})$,
   (b) Calculate the direction of search $\vec{d}(k)$ by solving the system of linear equations: $H(k)\,\vec{d}(k) = \text{grad}_{\vec{x}}\, J(\vec{x}(k))$, assuming that $H(k)$ is nonsingular, and go to the next step. If $H(k)$ is singular, set $\vec{d}(k) = \text{grad}_{\vec{x}}\, J(\vec{x}(k))$ and go to the next step.
   (c) Calculate a new sequence of decisions

$$\vec{x}(k+1) = \vec{x}(k) - \gamma\,\vec{d}(k). \tag{5.19}$$

   Set $k := k + 1$ and go to step 3.

Several remarks are in order concerning the above algorithm.

- Its potential applicability is wider than the COVID-19 decision making.
- In step 4 (b) the Hessian matrix is represented in a numerical way. One can imagine computations of the inverse of $H(\vec{x})$ in a symbolic way, but it is limited to small $N$ only.
- The Newton method, if convergent, is known to be rapidly (quadratically) convergent. One can try to extend its domain of convergence by selecting $\gamma$ that changes from pass to pass according to the line search paradigm. These approaches are known as damped Newton's methods. In such cases the calculations of $J$ are necessary. However, in the example reported in the next subsection, it was sufficient to use $\gamma = 0.95$.
- If in step 4 (b) Hessian matrix is singular or badly conditioned, one can calculate a new direction of search by solving the following system of linear equations:

$$(H(k) + \vartheta\, I_N)\,\vec{d}(k) = \text{grad}_{\vec{x}}\, J(\vec{x}(k)), \tag{5.20}$$

  where $I_N$ is an $N \times N$ unit matrix. This method of calculating the direction of search is known as the Levenberg-Marquardt algorithm. The reader is referred to [105], where suggestions concerning the choice of tuning parameter $\vartheta > 0$ can also be found. Observe that for $\vartheta \to 0+$ the solution of (5.20) approaches the one used in the Newton method, while for growing $\vartheta$ it is closer to the gradient direction of search.
- The improvements in (5.19) have the form of pass to pass learning, since the updates of $\vec{x}(k+1)$ depend on the decisions from the previous pass $\vec{x}(k)$ directly and through $\vec{d}(k)$.

### 5.3.3 *Testing Example—COVID-19 Mitigation*

The hybrid numerical-symbolic version of the Newton was applied for finding the minimizer of (5.16), where $s_N$ is recursively defined by (5.13), while the additional constraints have the form: for every pass $k$

$$x_{min} \leq x_n(k) \leq x_{max}, \quad n = 1, 2, \ldots, N, \tag{5.21}$$

$$|x_{n+1}(k) - x_n(k)| \leq \Delta x_{max}, \quad n = 1, 2, \ldots, N. \tag{5.22}$$

that were discussed in the previous section.

By projecting the direction of search $\vec{d}(k)$ onto boundaries of the parallelepiped defined by (5.21), which is easy in this case, one can assure that this group of constraints is fulfilled. Concerning the second group (5.22), the following rule was applied:

$$\text{if } |x_{n+1}(k) - x_n(k)| > \Delta x_{max}, \quad \text{then} \tag{5.23}$$

$$x_{n+1}(k) = x_n(k) + \Delta x_{max} \, \mathrm{sgn}[x_{n+1}(k) - x_n(k)].$$

Observe that when constraints (5.21) and (5.22) are present, then one cannot use the stopping condition (5.18), since the optimal solution with the non-vanishing gradient can be found at an active constraint. In the simulations reported below, the algorithm was stopped after 5 passes.

The learning processes were simulated for $N = 15$ and $N = 30$ days, assuming that the pandemic is close to expiring and our aim is to find a decision sequence of gradually removing social restrictions.

Each time the simulations were started from decision sequences generated at random according to the distribution that is uniform in $[0.5, 1.5]$. For such sequences constraints (5.22) may not be fulfilled, which makes the search for the optimal sequence more difficult.

The times (in seconds) of computations are reported in Table 5.2, where

- Ts is the time of computing $s_N$ as a function of $\vec{x}$,
- Tp is the time of preparing the gradient of $J(\vec{x})$ for further computations,
- Th is the time of computing the Hessian matrix with elements being functions of $\vec{x}$,
- Tpass is the computational time needed for calculating the update of the sequence of decisions for one pass.

Notice that computations of $s_N$, $J$ and $H$ as functions of $\vec{x}$ are performed only once for fixed numerical values of $r$, $K$ and $s_{init}$.

Figures 5.5 and 5.7 illustrate that the hybrid Newton method is convergent in two learning passes. Figures 5.6 and 5.8 confirm this conclusion.

**Table 5.2** Timing (in sec.) of computing Ts—the state $s_N$, Td—the derivatives of $J(\vec{x})$ and Th—the Hessian matrix as well as time Tpass of performing the update for one pass (see the text for more explanations)

| | Ts | Td | Th | Tpass |
|---|---|---|---|---|
| $N = 15$ | 14 | 2.2 | 10 | 12 |
| $N = 30$ | 3323 | 96.3 | 807 | 1115 |

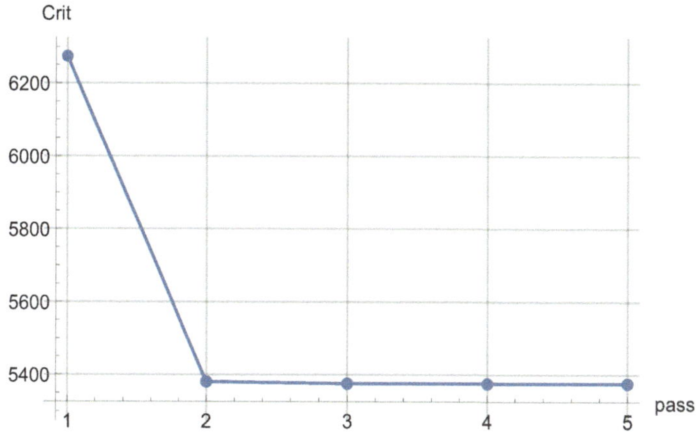


**Fig. 5.5** The goal function $J$ in subsequent iterations of learning by the hybrid version of the Newton method for $N = 15$

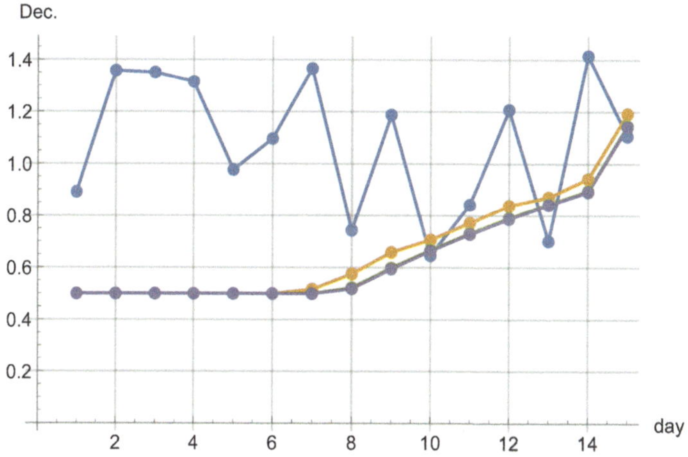


**Fig. 5.6** The improvements of the decision sequence in subsequent passes of learning by the hybrid Newton method for $N = 15$

Conclusions concerning practical decisions on mitigating the spread of COVID-19 are certainly outside the scope of this book. However, it is worth observing that the optimal decision sequences shown in Figs. 5.6 and 5.8 suggest maintaining social restrictions at a high level ($x_n$'s close to 0.5) for about half of the period and then to remove them gradually ($x_n$'s growing to about 1.2–1.5.). Notice that the same values of $x_n$'s can be obtained in practice by different combinations of the actions listed at the beginning of this chapter. Closer analysis of the relationships between $x_n$'s and these actions is outside the scope of this book because of the lack of data that would be necessary for identifying such dependencies.

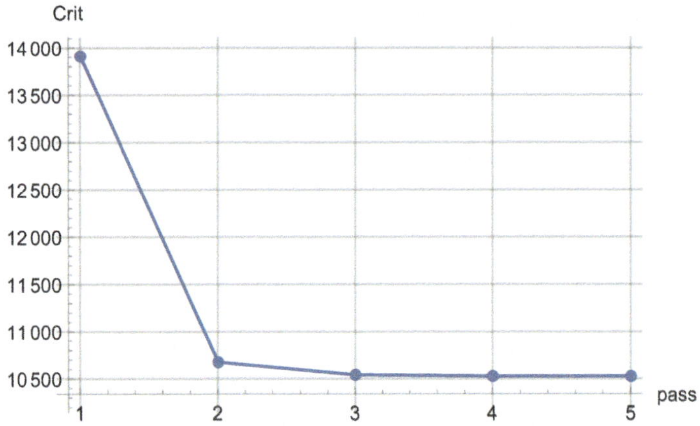


**Fig. 5.7** The goal function $J$ in subsequent passes of learning by the hybrid version of the Newton method for $N = 30$

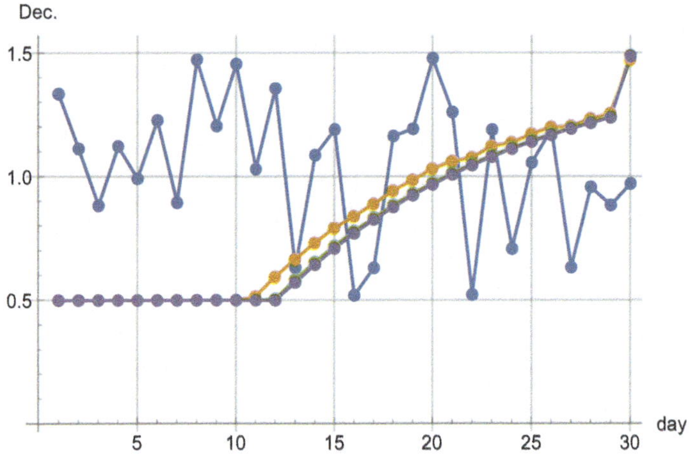


**Fig. 5.8** The improvements of the decision sequence in subsequent passes of learning by the hybrid Newton method for $N = 30$

Nevertheless, the following facts are qualitatively convincing that the learning of optimal decisions can be beneficial. To this end, assume that the sequence $x_n = 1$ is applied all the time. Then, the predicted number of infected people would be 5614 for $N = 15$ and 11265 for $N = 30$. If the optimal decision sequences (shown in Figs. 5.6 and 5.8) are applied, then the predicted number of infected would be 4381 for $N = 15$ and 8382 for $N = 30$. The reduction is remarkable, taking into account that the goal function $J$ contains the costs of too excessive social restrictions.

In the above case studies a purely model-based approach for learning was applied, since experimental running passes in real life is not possible. However, when more and more countries are closer to suppressing the epidemic, one can improve its model and re-run computations.

The analysis of Table 5.2 and Figs. 5.5, 5.7 indicates that the hybrid version of the Newton algorithm can be useful for computing optimal sequences of decisions for a moderate horizon. However, the computational time grows very rapidly with $N$ and for larger $N$ it is desirable to apply purely numerical algorithms, e.g. such as the differential evolution method with population filter (see Chap. 4). This algorithm is tested in the next section using the same COVID-19 example.

## 5.4 Learning Decisions by Differential Evolution with Population filter

The differential evolution with population filter is an optimization method that can be used to solve the problem stated previously. Its robustness and ability to find a global optimum can prove usefull.

For the sake of clarity let us now show the complete optimization problem:

$$\min_{x} J(\vec{x}) = s_N + \lambda \sum_{n=0}^{N} \gamma^n (x_{max} - x_{n-1})^2 , \tag{5.24}$$

with respect to

$$x_{min} \le x_n \le x_{max}, \quad |x_{n+1} - x_n| \le \Delta x_{max} . \tag{5.25}$$

The parameters are presented in Table 5.3

### 5.4.1 Solving the Optimization Problem

Firstly, simulations were carried out multiple times. The results can be seen in Fig. 5.9. The 300 iteration limit seems viable and was used in subsequent calculations. Note that practically all solutions are near the same at the end of calculations. The calculations were carried out for $N = 15$ as in the previous section.

**Table 5.3** Parameters of the COVID optimization problem

| $S_0$ | $S_0$ |
|---|---|
| $K$ | 40000 |
| $\lambda$ | $1.25 \cdot 10^{-3}$ |
| $\gamma$ | 1.08 |
| $x_{min}$ | 0.5 |
| $x_{max}$ | 1.5 |
| $\Delta x$ | 0.25 |

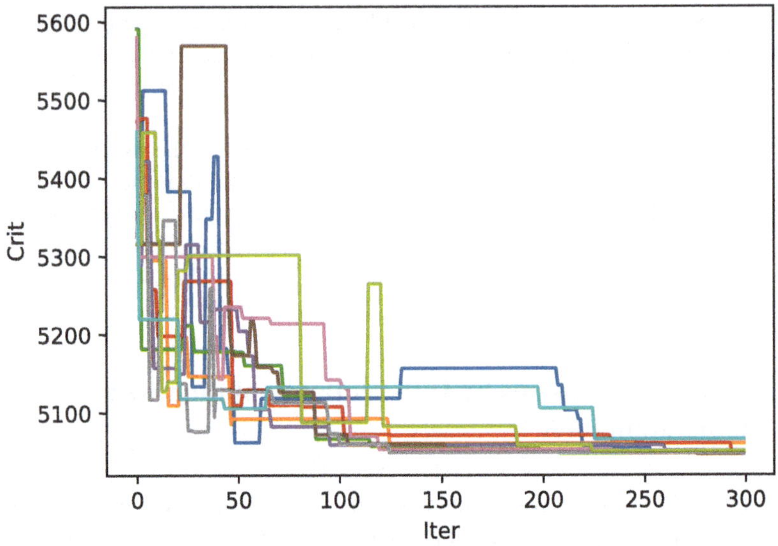


**Fig. 5.9** Multirun result in subsequent passes of learning from differential evolution with population filter

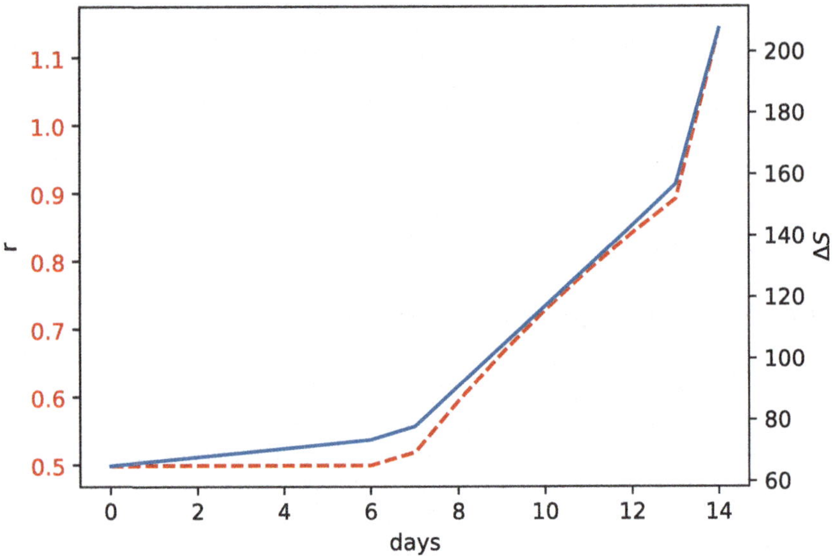


**Fig. 5.10** Daily cases (right axis) and $r$ (left axis, dashed line) obtained for $N = 15$

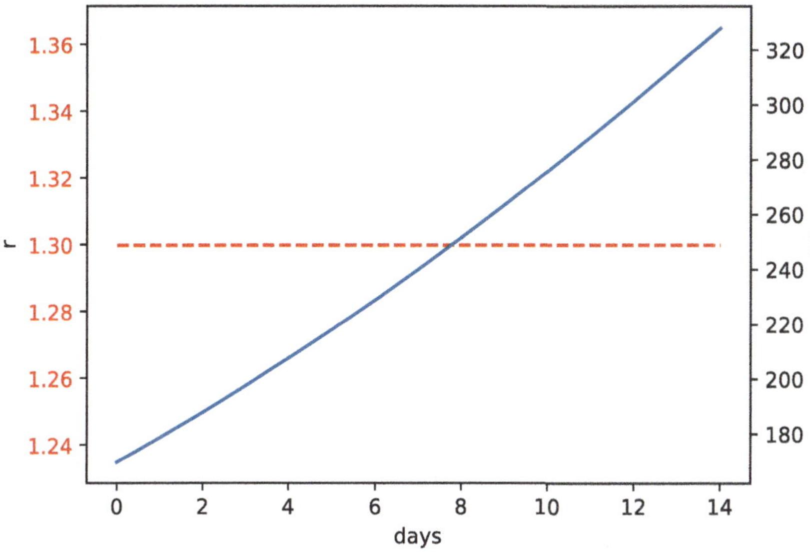


**Fig. 5.11** Daily cases (right axis) with constant $r = 1.3$ (left axis, dashed line) obtained for $N = 15$

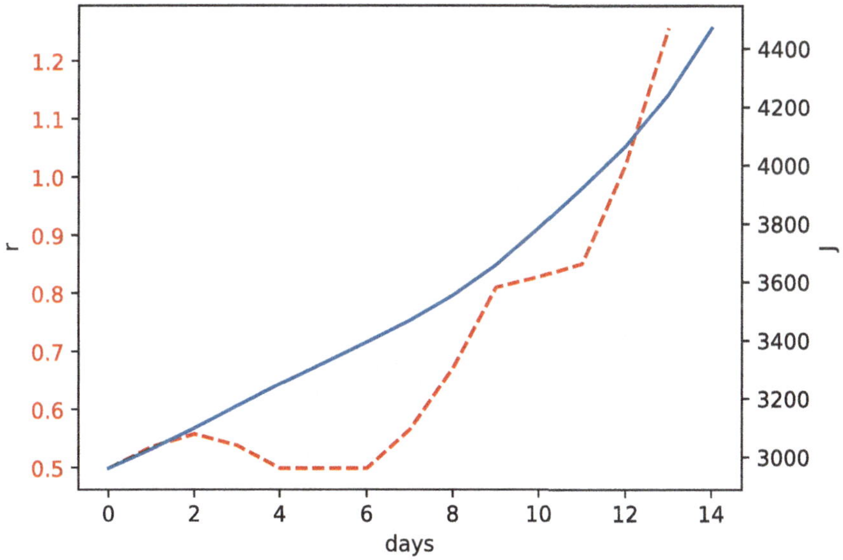


**Fig. 5.12** Value of the goal function (right axis) and $r$ (left axis, dashed line) obtained for $N = 15$

The results obtained for $N = 30$ are $J = 10541$, comparable with symbolic calculations. The resulting restrictions $r$ are presented in Fig. 5.13. Other cases are in Figs. 5.12 and 5.13.

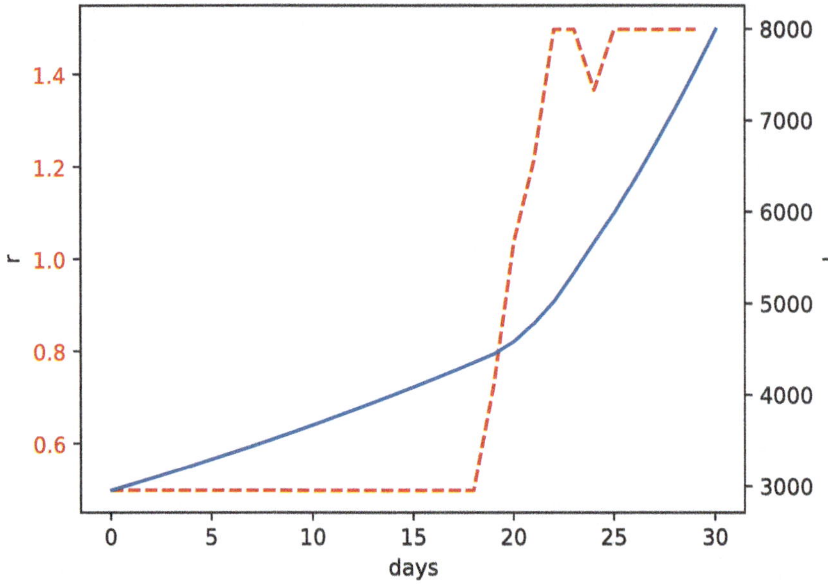


**Fig. 5.13** Value of the goal function (right axis) and $r$ (left axis, dashed line) obtained for $N = 30$

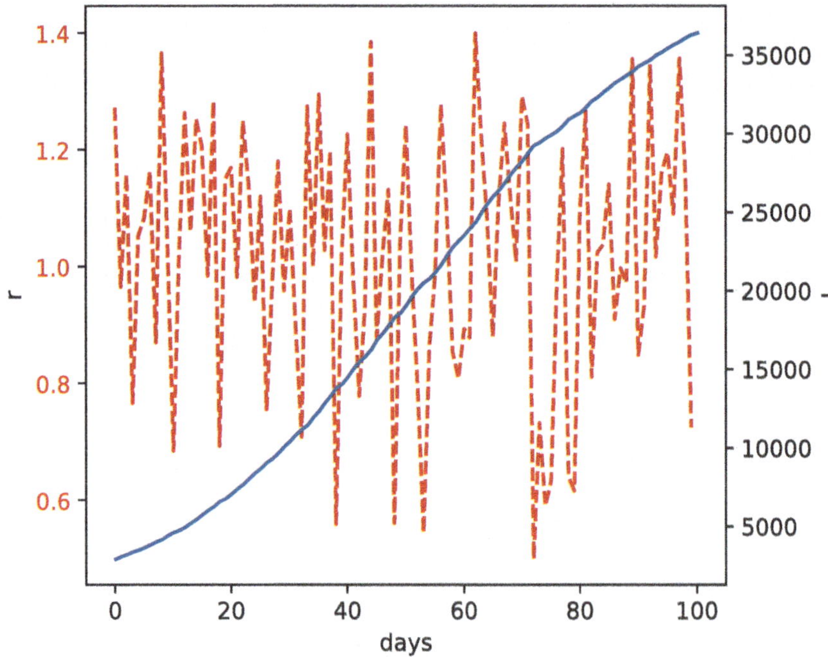


**Fig. 5.14** Value of the goal function (right axis) and $r$ (left axis, dashed line) obtained for $N = 100$

**Table 5.4** Timing (in sec.) of computing using differential evolution with population filter

| | T[s] |
|---|---|
| $N = 15$ | 2.8 |
| $N = 30$ | 4 |
| $N = 60$ | 6.3 |
| $N = 100$ | 8.7 |

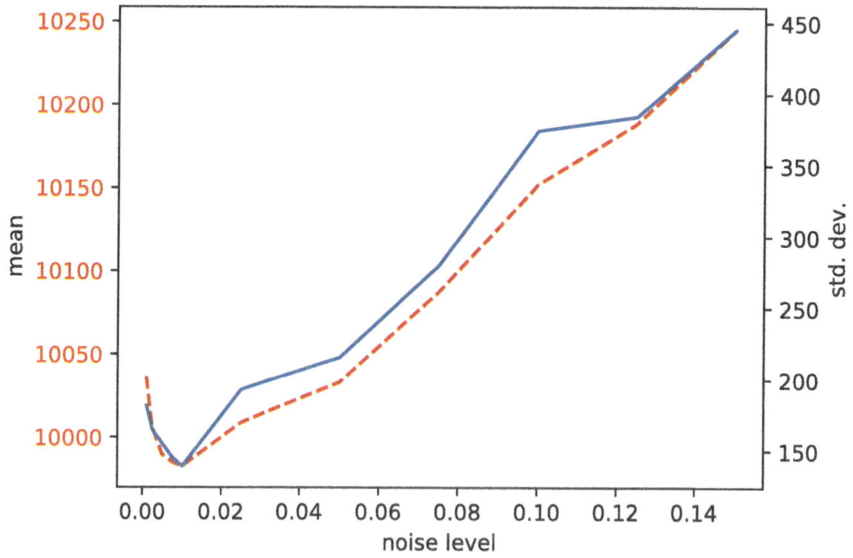


**Fig. 5.15** Mean value of the goal function (left axis, dashed line) and standard deviation (right axis) obtained for $N = 30$ with a different noise level

The daily case number increase can be seen in Fig. 5.10 and should be compared with constant $r = 1.3$ in Fig. 5.11 (see Table 5.4 for the timings for larger $N$).

### *5.4.2 Robustness of Differential Evolution*

Differential evolution is a very robust method. It could deliver a satisfying solution even with an uncertain or stochastic goal function. This feature can be helpful (Fig. 5.14).

We must realize that model (5.24) is dependent on a few parameters and independent of any other factors. The parameters are estimates and are uncertain, at least to some degree. Also the impact of the restrictions $r_i$ is treated as changing with time only—the longer restrictions last, the higher their impact. We do not take other factors like weather into account.

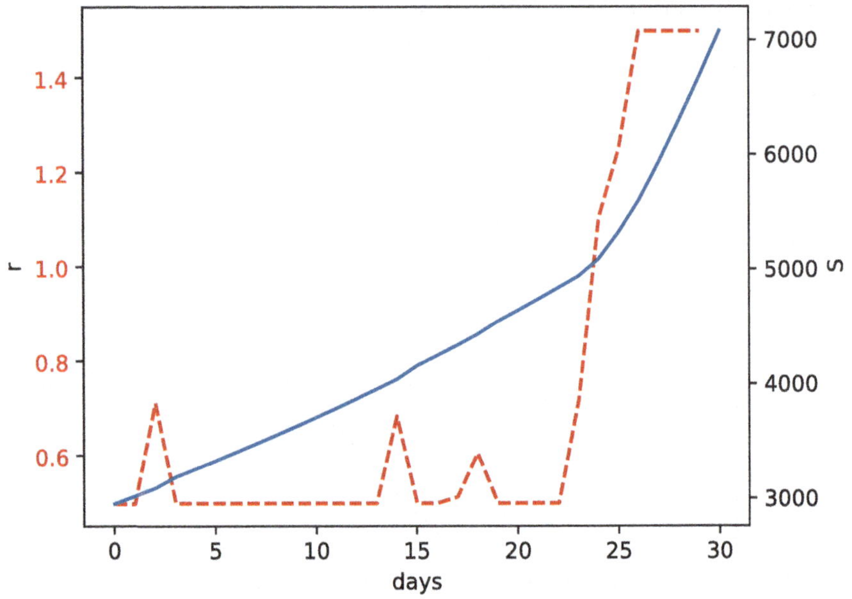


**Fig. 5.16** Number of infected (right axis) and $r$ (left axis, dashed line) obtained for $N = 30$ with 10% randomness in the goal function

The simplest solution is to add random noise to the goal function. Many random distributions can be used. Here we use the uniform distribution.

$$\min_{x} J(\vec{x}) = s_N + \lambda \sum_{n=0}^{N} \gamma^n (x_{max} - x_{n-1})^2 + \eta \cdot s_N \cdot U[-1, 1], \tag{5.26}$$

with respect to

$$x_{min} \leq x_n \leq x_{max}, \quad |x_{n+1} - x_n| \leq \Delta x_{max} . \tag{5.27}$$

where $\eta \in [0, 1]$ is the noise level.

The resulting restrictions for $N = 30, \eta = 0.05$ can be seen in Fig. 5.16. The calculations were also carried out for different noise levels and the resulting goal function (the final result calculated without the noise) can be seen in Fig. 5.15. The results are the average from 1000 runs for each noise level.

To conclude the differential evolution simulations, we should note that the ability to calculate decisions in a very long period is only a computational ability. The model is much less accurate in the long run. The more important factor making differential evolution a better tool is its robustness.

Differential evolution is a global optimization method. In this case numerous terms in the COVID-19 polynomial model can suggest that the problem is multimodal and such a method is indeed required.

# Chapter 6
# Stochastic Gradient in Learning Decision Sequences

In this chapter the known, classic and more recent approaches to learning decisions and decision sequences that are based on various versions are reviewed. They are presented in a unified manner in order to compare them from the point of view of their applicability to learn relatively long decision sequences of the length, say, of dozens or hundreds. The following algorithms are discussed:

1. the classic Kiefer–Wolfowitz stochastic approximation, which is a model-free approach,
2. a random, simultaneous perturbations version of the above,
3. the so-called response surface methodology that can be viewed as a model-inspired approach to estimating the gradient.

Their common features are:

- the necessity of process-algorithm interaction—the need for observations (measurements) to evaluate the impact of each decision sequence on a process (a system),
- applicability to processes that are repeatable,
- improvements of a decision sequence as a whole entity.

The reviewed approaches are ordered from model-free to model-inspired ones.

The focus point is in learning the whole decision sequence of a repetitive process from pass to pass (from run to run or from trial to trial). This is in contrast to the classic approach in automatic control that usually corrects errors along each pass only, without transferring information between passes. On the other hand, as already mentioned in the Introduction, at least from the last two decades of the 20th century, the iterative learning control (ILC) stream of research has been intensively developing. Its main message is learning from pass to pass, which is an inspiration for the approach considered in the next chapter.



W. Rafajłowicz, *Learning Decision Sequences For Repetitive Processes—Selected Algorithms*, Studies in Systems, Decision and Control 401,
https://doi.org/10.1007/978-3-030-88396-6_6

## 6.1 Model-Free Classic Approach—Stochastic Approximation Revisited

In Chap. 3 stochastic approximation has already been discussed from a general viewpoint of learning algorithms and its relationships to other learning approaches. In this section, we provide a closer look at its properties. We put emphasis on its modifications that are able to learn between passes, but require a much smaller number of trials (process runs or simulations) to estimate a descent direction.

The original, Robbins–Monroe version of the stochastic approximation procedure [146] was dedicated to finding zeros of a regression function(s). This procedure was deeply investigated and generalized in many directions, including averaging of the iterates [121, 158], convergence in the Hilbert spaces [178], the rate of convergence [122], the law of iterated logarithm [78] (and references therein for earlier results on the convergence rates), asynchronous version: [19, 141], robust version [102] and many others (see e.g. [29, 81, 82, 175, 181] for comprehensive expositions of these subjects).

### 6.1.1 *Stochastic Approximation—Problem Statement*

For our purposes, its version developed by the Kiefer and Wolfowitz [69] and its further variants mentioned below is more relevant. We keep the historical term *stochastic approximation*, although in this case the name *stochastic gradient descent* would better reflect the essence of this approach. The Kiefer–Wolfowitz stochastic approximation algorithm (K-WSAA) was designed to learn a location of a (local) minimum $\vec{x}^* \in \mathrm{R}^{d_y}$ of a regression function that is defined as follows:

$$\mu(\vec{x}) = \mathbb{E}_y\left(y|\vec{X} = \vec{x}\right) = \int_{\mathrm{R}^{d_y}} y\, f_y(y|\vec{x})\, dy\,, \tag{6.1}$$

where $\mathbb{E}_y\left(y|\vec{X} = \vec{x}\right)$ denotes the conditional expectation of $y$, given $\vec{x}$ (see Chap. 2 for more explanations). In other words, the aim of the K-WSAA is to learn

$$\vec{x}^* = \arg\min_{\vec{x}\in\mathrm{R}^{d_y}} \mu(\vec{x}), \tag{6.2}$$

assuming that any local minimum of $\mu(\vec{x})$ solves the problem and this function is not known and $\mu(\vec{x})$ cannot be directly measured. Instead, for each fixed $\vec{x}$ one is able to get observation $y(\vec{x})$ of a random variable $Y(\vec{x})$ that has $f_y(y|\vec{x})$ as the conditional p.d.f. with the expectation $\mu(\vec{x})$.

In practice, this means that if at $k$th iteration of learning a decision sequence $\vec{x}_k$ is applied to a process (to a system), then one obtains $y_k = y(\vec{x}_k)$ as its output (reaction, quality index). It is interpreted as a realization of r.v. $Y(\vec{x}_k)$, having $\mu(\vec{x}_k)$ as its conditional expectation.

If function $\mu$ would be known and continuously differentiable, then one could find $\vec{x}^*$ by solving the following set of equations: $\nabla\mu(\vec{x}) = \bar{0}$, where $\nabla$ stands for the gradient of $\mu(\vec{x})$ with respect to $\vec{x}$. Kiefer and Wolfowitz proposed estimating $\nabla\mu(\vec{x})$ at $\vec{x} = \vec{x}_k$ by vector $\vec{d}_k$ defined as follows:

$$\vec{d}_k = \frac{1}{2\,c_k} \begin{bmatrix} y(\vec{x}_k + c_k\,\vec{e}_1) - y(\vec{x}_k - c_k\,\vec{e}_1) \\ y(\vec{x}_k + c_k\,\vec{e}_2) - y(\vec{x}_k - c_k\,\vec{e}_2) \\ \dots \\ y(\vec{x}_k + c_k\,\vec{e}_N) - y(\vec{x}_k - c_k\,\vec{e}_N) \end{bmatrix}, \tag{6.3}$$

where $c_k > 0,\ k = 1,\ 2, \dots$ is a sequence convergent to zero at an appropriate rate, while $\vec{e}_i$ is $i$th unit vector $\underbrace{[0\ 0, \dots,\ 1,\ 0, \dots,\ 0]^T}_{i\text{th}}$ of the standard basis, $i = 1, 2, \dots, N$.

### 6.1.2 The Kiefer–Wolfowitz Algorithm

The classic K-WSAA runs as follows: set $k = 0$ and select a starting point $\vec{x}_0$, then iterate

$$\vec{x}_{k+1} = \vec{x}_k - \alpha_k\,\vec{d}_k \equiv \vec{x}_{k+1} = (1 - \alpha_k)\,\vec{x}_k + \alpha_k\,(\vec{x}_k - \vec{d}_k), \tag{6.4}$$

where $\alpha_k > 0,\ k = 0,\ 1, \dots$ is a sequence of numbers. Its choice is discussed later. The second, equivalent version in (6.4) is in the form typical for an exponentially weighted moving average (EWMA). Notice that $y$'s in (6.3) are realizations of r.v.'s, which implies that also $\vec{d}_k$'s and $\vec{x}_k$ are random. Hence, a stopping condition of this and other stochastic gradient algorithms should not be based on the magnitude of one $||\vec{d}_k||$, but rather on the averages of $||\vec{d}_k||$'s from several iterations or directly on attaining the maximum of the admissible number of iterations.

Also possible convergence of $\vec{x}_k$'s to $\vec{x}^*$ should be considered in a probabilistic sense. Fortunately, several theorems, providing sufficient conditions for convergence $\vec{x}_k$'s to $\vec{x}^*$ with probability one and in the mean squared error (MSE) sense, have been proved. In the most of them the following conditions are imposed on the simultaneous, asymptotic (as $k \to \infty$) choice of the step-size $\alpha_k \to 0$ and on $c_k \to 0$:

$$\sum_{k=1}^{\infty} \alpha_k = \infty, \quad \sum_{k=1}^{\infty} \alpha_k\,c_k < \infty, \quad \sum_{k=1}^{\infty} \alpha_k^2\,c_k^{-2} < \infty. \tag{6.5}$$

These conditions are usually interpreted as follows:

- condition $c_k \to 0$ ensures that $\vec{d}_k$ is asymptotically unbiased for the gradient of $\mu$,
- condition $\alpha_k \to 0$ and the first requirement in (6.5) say that the step size should be convergent to zero, but not in a too fast way,

- additionally, as the second condition in (6.5) implies, the rate of convergence of $c_k$'s should be sufficiently fast so as to force the series $\alpha_k\, c_k$ to be convergent,
- however, this rate cannot be faster than those of $\alpha_k$'s as the third condition in (6.5) requires.

Selecting $\alpha_k = \varsigma_1/n^a$, $c_k = \varsigma_2/n^b$, where $\varsigma_1 > 0$, $\varsigma_2 > 0$ are preselected constants, one can fulfill conditions (6.5) by choosing the exponents $a > 0$, $b > 0$ in such a way that

$$0 < a \leq 1, \quad a + b > 1, \quad a - b > 1/2. \tag{6.6}$$

These conditions hold in the admissible triangle $\frac{3}{4} < a \leq 1$ and $1 - a < b < \frac{1}{2}(2a - 1)$, as sketched in Fig. 6.1. In addition to (6.5), a number of regularity conditions are imposed on $\mu(\vec{x})$ and its gradient. A typical set of them includes:

$$\mathrm{Var}(Y(\vec{x})) < const, \quad \mu(\vec{x}) \text{ is strictly convex} \tag{6.7}$$

Under (6.5) and (6.7), extended by even more technical conditions, the convergence of $\vec{x}_k$ to $\vec{x}^*$ with probability one and in the MSE sense is proved by Blum [12] in the univariate case. The proofs are—in most cases—based on martingales (see, e.g., [13]) or on constructing an appropriate set of ordinary differential equations (ODEs) that approximates a deterministic part of the behavior of (6.4), further denoted as $\tilde{x}(t)$. Then, the stability of these ODEs at $\vec{x}^*$ is investigated (see e.g. [82]), assuming that the gradient of $\mu(\vec{x})$ at $\vec{x}^*$ is the vector of zeros.

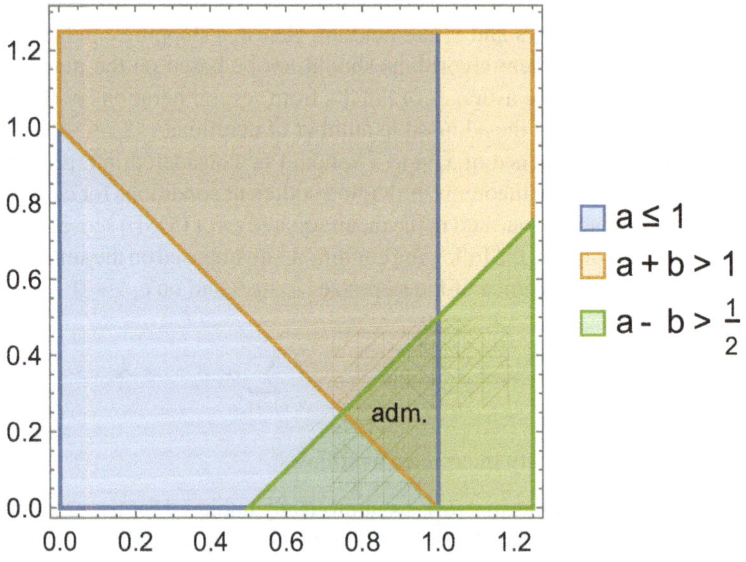


**Fig. 6.1** The admissible triangle for the exponent pairs $(a,\ b)$, defined by (6.6)

In a typical case the ODEs have the form:

$$\frac{d\,\tilde{x}(t)}{d\,t} = \nabla_{\vec{x}}\,\mu(\vec{x})\Big|_{\vec{x}=\tilde{x}(t)}, \quad \tilde{x}(0) = \tilde{x}_0. \tag{6.8}$$

The Lipschitz condition imposed on $\nabla_{\vec{x}}\,\mu(\vec{x})$ is sufficient for the existence and uniqueness of the solution of (6.8).

Let us assume that in a certain vicinity $\mathcal{O}(\vec{x}^*)$ of $\vec{x}^*$ the following condition holds: for every $\vec{x} \neq \vec{x}^*$

$$(\vec{x} - \vec{x}^*)^T\,\nabla_{\vec{x}}\,\mu(\vec{x}) < 0. \tag{6.9}$$

Selecting the Liapunov function as $V(\tilde{x}) = ||\vec{x} - \vec{x}^*||^2/2$ and differentiating $V(\tilde{x}(t))$ along the solution of (6.8), we obtain, for every $t > 0$,

$$\frac{d\,V(\tilde{x}(t))}{d\,t} = (\tilde{x}(t) - \vec{x}^*)^T\,\nabla_{\vec{x}}\,\mu(\vec{x})\Big|_{\vec{x}=\tilde{x}(t)}, \quad \tilde{x}(0) = \tilde{x}_0. \tag{6.10}$$

Denote by $\mathcal{V}(\rho)$ the family of sets

$$\mathcal{V}(\rho,\,\vec{x}^*) = \{\vec{x} : V(\vec{x} - \vec{x}^*) \leq \rho\}, \quad \rho > 0 \tag{6.11}$$

and let $\rho_{max}$ be the largest $\rho$ such that

$$\mathcal{V}(\rho,\,\vec{x}^*) \subseteq \mathrm{cl}[\mathcal{O}(\vec{x}^*)], \tag{6.12}$$

where cl[.] denotes the closure of a set in the brackets. We admit $\rho_{max} = \infty$, if $\mathcal{O}(\vec{x}^*)$ is the whole space. Observe that by (6.9), if $\tilde{x}_0 \in \mathcal{V}(\rho_{max},\,\vec{x}^*)$, then also $\tilde{x}(t) \in \mathcal{V}(\rho_{max},\,\vec{x}^*)$ for all $t > 0$. Additionally, again by (6.9) and by the second Liapunov method, $\lim_{t\to\infty} ||\tilde{x}(t) - \vec{x}^*|| = 0$, which means that the deterministic part of the approximation of $\vec{x}_k$'s converges to $\vec{x}^*$. The full proof of the convergence of the random part of $\vec{x}_k$'s to zero in the MSE sense would require the analysis of stochastic differential equations, which depends on assumptions concerning random fluctuations of $Y$'s that are difficult to be verified and therefore it is omitted.

### 6.1.3 *Modifications of the Kiefer–Wolfowitz Algorithm*

Since the 1950s a number of modifications of the Kiefer–Wolfowitz algorithm have been proposed. Most of them are directed to reduction of the main weakness of the original K-WSAA, namely, its relatively slow convergence (in practice, frequently in hundreds or even thousands of iterations). These modifications include:

1. An adaptive choice of the step length $\alpha_k$, i.e., instead of selecting a deterministic sequence, for which conditions (6.5) hold, $\alpha_k$ is selected according to rules depending on whether previous step(s) was successful or not.

2. Attempts at minimizing the goal function (observed with random errors) in direction $\vec{d}_k$, by conducting experiments. For our purposes, this generalization has limited applicability, due to the necessity of running the process additionally at least several times per iteration.
3. Adding the momentum to (6.4), i.e., iterate the following:

$$\vec{x}_{k+1} = \vec{x}_k - \alpha_k \vec{d}_k + \gamma_k (\vec{x}_k - \vec{x}_{k-1}), \tag{6.13}$$

   where $\gamma_k$'s is, in general, a preselected sequence, but the choice $\gamma_k = \gamma > 0$ for all $k$ has its advantages (see [22], Eq. (7.2)).
4. A seemingly minor, but important modification of (6.13) was proposed by Nesterov (see [103] and [22], Eqs. (7.4), (7.5)). Its essence is to replace $\vec{d}_k$ in (6.13) by the gradient estimate centered around $\vec{x}_k + \gamma_k (\vec{x}_k - \vec{x}_{k-1})$. This modification is appealing, since—as mentioned in [22] (Sect. 7.2)—if a goal function is convex and has a continuously differentiable gradient, which is Lipschitz continuous, then the convergence rate of the Nesterov algorithm is of the order $k^{-2}$, assuming that $\alpha_k$ is a constant and $\gamma_k$ converges to 1 from below in a monotonically increasing way. Comparison of this rate with $O(k^{-1})$ that is typical for the steepest descent, explains the name *accelerated gradient method*. One may hope that this desirable rate is retained also in the presence of large observation errors.
5. Averaging previous $\vec{x}_k$'s, since the K-WSAA behaves, to some extent, as a typical gradient algorithm, for which this kind of averaging is in certain cases beneficial.
6. Applying the averaging of stochastic gradients $\vec{d}_k$ 's, which leads to estimates of the true gradient with smaller variances of the components.
7. Modifying the search directions $\vec{d}_k$ 's using the guidelines of deterministic search algorithms. In particular,

   - applying conjugate gradient algorithms, substituting $\vec{d}_k$ 's as the gradients,
   - approximating the hessian matrix of the second derivatives of the goal function, denoted further by $\hat{H}_k$, in the spirit of the Newton-Raphson method, i.e.,

$$\hat{H}_k = \sum_{i=1}^{k} \vec{d}_k \, \vec{d}_k^T \tag{6.14}$$

     and running the iterations as follows: run at least the first $N$ iterations according to (6.4), so as to obtain $\hat{H}_k$ nonsingular, and then iterate

$$\vec{x}_{k+1} = \vec{x}_k - \alpha_k \hat{H}_k^{-1} \vec{d}_k, \quad k = (N+1), (N+2), \ldots \tag{6.15}$$

   - as above, but running the iterations in the Levenberg–Marquardt manner

$$\vec{x}_{k+1} = \vec{x}_k - \alpha_k (\hat{H}_k + \lambda I_N)^{-1} \vec{d}_k, \quad k = (N+1), (N+2), \ldots,$$

     where $\lambda > 0$ should be separately chosen,

- approximating $\hat{H}_k^{-1}$ directly (without the explicit matrix inversion) in the spirit of the BFGS[1] method and using it in (6.15), together with the attempts to optimize the step length.

8. Updating only a part of $\vec{x}_k$ at each iteration—known as asynchronous stochastic approximation.

The next stream of generalizations of the K-WSAA is devoted to optimization problems with constraints imposed on $\vec{x}$.

### 6.1.4 *Generalizations of the K-WSAA for Handling Constraints*

In the simplest case, when separate constraints are imposed on the elements of $\vec{x}$ as lower bounds $a_i$ and upper bounds $b_i$, $a_i < b_i$, $i = 1, 2, \ldots, N$, one can modify the K-WSAA as follows:

$$\vec{x}_{k+1} = \Pi[\vec{x}_k - \alpha_k \vec{d}_k], \; k = 0, 1, \ldots, \tag{6.16}$$

where $\Pi[.]$ reduces the elements of a vector in the brackets in such a way that they stay within these bounds.

The orthogonal projection of the search direction $\vec{x}_k - \alpha_k \vec{d}_k$ onto a set of linear (in $\vec{x}$) constraints is still possible, since the result of the projection can be explicitly computed. When constraints are nonlinear, the result of the projection onto the plane tangent to the set of admissible solutions may fail to stay within it.

Nonlinear constraints can be approximately taken into account also by adding a penalty (inner or outer) to the observations of the goal function. The projection and the penalty function approaches can be recommended mainly when one can directly verify whether $\vec{x}$ fulfills the constraints or not. If constraints are given implicitly, as it happens when they are imposed on states of dynamical systems, then the idea of the filter of solutions (see Chap. 4) can be useful.

The Kiefer–Wolfowitz method played an important role in the development of decision learning algorithms. However, in its original form, this method is not useful for long decision sequences $\vec{x}$. The reason is that one has to run a process to be optimized $2N$ times at each iteration[2] to estimate $\vec{d}_k$, visible from (6.3), which may be prohibitively expensive or time-consuming for large $N = dim(\vec{x})$.

This is also the reason that above we use the traditional term *iteration* and the notation $\vec{x}_k$ for one step of the K-WSAA.

---

[1] Broyden–Fletcher–Goldfarb–Shanno.

[2] Furthermore, the reduction of this burden from $2N$ to $N$, by using one-sided differences, does not help too much if $N$ is large.

## 6.2 Random, Simultaneous Perturbations for Estimation Gradient at Low Cost

The remedy for this drawback of the K-WSAA was proposed a long time ago. The idea is to add a small number of random perturbations to the whole sequence $\vec{x}_k$ instead of (6.3). In [75] (see also [76]) it was proposed to add unit vectors that are uniformly distributed on the unit sphere or unit cube. If these random vectors are mutually independent and a number of technical conditions hold, then it is possible to prove the convergence of the learning process (see [75] for rigorous proof).

Another variant of adding a small number (two or even one, under additional assumptions) random perturbations was proposed in [166, 167] and the convergence rate is investigated in [30]. At present, this approach appears under different names, e.g. a Kiefer–Wolfowitz Algorithm with randomized differences, simultaneous perturbation method, simultaneous perturbation gradient approximation and others.

### 6.2.1 *The Idea of Simultaneous Random Perturbations*

It is convenient to explain the essence of this variant for a bivariate, continuously differentiable function $\phi(x_1, x_2)$, say. In order to estimate its partial derivatives at $(x_1, x_2)$, let us add a random vector $[\Delta_1, \Delta_2]^T]$. Then, by the Taylor expansion, we obtain:

$$\phi(x_1 + c'\Delta_1, x_2 + c''\Delta_2) \approx \phi(x_1, x_2) + c'\Delta_1 \frac{\partial\phi(x_1, x_2)}{\partial x_1} + c''\Delta_2 \frac{\partial\phi(x_1, x_2)}{\partial x_2},$$

where $c' > 0$ and $c'' > 0$ play the role of the step sizes. Let us assume that

1. $\mathbb{E}[\Delta_1] = 0, \mathbb{E}[\Delta_2] = 0$,
2. $\Delta_1$ and $\Delta_2$ are stochastically independent,
3. $\mathbb{E}[1/\Delta_1]$ and $\mathbb{E}[1/\Delta_2]$ exist and they are finite.

Observe that the last assumption excludes the Gaussian distribution from considerations, but one can select, e.g. Bernouli random variables, taking values $\pm 1$ with the probability $1/2$, as $\Delta_1$ and $\Delta_2$.

As an estimate of $\frac{\partial\phi(x_1,x_2)}{\partial x_1}$ consider $D_1$ that is defined as follows:

$$D_1 = \frac{\phi(x_1 + c'\Delta_1, x_2 + c''\Delta_2) - \phi(x_1, x_2)}{c'\Delta_1} \tag{6.17}$$

and analogously for $\frac{\partial\phi(x_1,x_2)}{\partial x_2}$:

$$D_2 = \frac{\phi(x_1 + c'\Delta_1, x_2 + c''\Delta_2) - \phi(x_1, x_2)}{c''\Delta_2}. \tag{6.18}$$

Notice that the nominator in (6.17) and (6.18) are intentionally the same. The difference between these two expressions is only in their denominators. To convince ourselves that $D_1$ and $D_2$ are (approximately[3]) unbiased estimators for $\frac{\partial\phi(x_1,x_2)}{\partial x_1}$ and $\frac{\partial\phi(x_1,x_2)}{\partial x_2}$, respectively, let us express them, using (6.21) as the equality, as follows:

$$D_1 = \frac{\partial\phi(x_1, x_2)}{\partial x_1} + \frac{c''\Delta_2}{c'\Delta_1}\frac{\partial\phi(x_1, x_2)}{\partial x_2}, \tag{6.19}$$

$$D_2 = \frac{c'\Delta_1}{c''\Delta_2}\frac{\partial\phi(x_1, x_2)}{\partial x_1} + \frac{\partial\phi(x_1, x_2)}{\partial x_2}. \tag{6.20}$$

Hence, $\mathbb{E}(D_1) = \frac{\partial\phi(x_1,x_2)}{\partial x_1}$, since by the above assumptions 1.-3. we obtain:

$$\mathbb{E}\left[\frac{\Delta_2}{\Delta_1}\right] = \mathbb{E}(\Delta_2)\,\mathbb{E}\left[\frac{1}{\Delta_1}\right] = 0 \tag{6.21}$$

and analogously for $\mathbb{E}(D_2)$.

Notice that the above considerations hold also when the two values of $\phi$ are observed with random noise, $\epsilon$ and $\epsilon_{step}$, say, with zero mean, if we additionally assume that $\epsilon$ and $\epsilon_{step}$ and $\Delta_1$, $\Delta_2$ are mutually independent. Indeed, in this case additional random summands:

$$(\epsilon_{step} - \epsilon)/(c'\Delta_1) \text{ and } (\epsilon_{step} - \epsilon)/(c''\Delta_2)$$

appear in (6.19) and (6.20), respectively, but they also have zero mean values.

### 6.2.2 *Simultaneous Perturbation Algorithm for Decision Learning (SPADL)*

As in the original Kiefer–Wolfowitz procedure, one can apply the symmetric differences to estimate $\frac{\partial\phi(x_1,x_2)}{\partial x_1}$ and $\frac{\partial\phi(x_1,x_2)}{\partial x_2}$, but at the expense of running a process twice at each iteration. From the point of view of optimizing the process, the on-line, one-sided version (6.17), (6.18) seems to be preferable, unless a model of the process is not available, because it allows for the process improvement after each pass, also for larger $N$, as described below. At this point we return to our standard notation $\vec{x}(k)$ for the sequence of decisions applied to a process at its $k$th pass.

**The SPADL Algorithm**

**Preparations:** Select initial guess $\vec{x}(init)$ and sequences $\alpha_k$' and $c_k$'s for which conditions (6.5) hold. Choose also a probability distribution for elements $\Delta_j(k)$, $j = 1, 2, \ldots, N$ of random perturbation vectors $\vec{\Delta}(k)$,

[3] The approximation results from applying the Taylor expansion has the same accuracy as the remainder of Taylor's series.

$k = 0, 1, \ldots$ in such a way that the following conditions hold for $j = 1, 2, \ldots, N$ and for all $k \geq 0$

- $\mathbb{E}[\Delta_j(k)] = 0, \quad \mathbb{E}[1/\Delta_j^l(k)] < \infty, \ l = 1, 2,$
- all r.v.'s $\{\Delta_j(k)\}$ are mutually independent.

 Select also a stopping condition, in the simplest case by bounding the number of updates by $k_{max} > 1$.

**Initialization:** Apply decision sequence $\vec{x}(init)$ to the process, acquire (observe) its output (response) $y(init)$ and store it. Draw random vector $\vec{\Delta}(0)$ with elements $\Delta_i(0)$, $i = 1, 2, \ldots, N$ and form decision sequence:

$$\vec{x}(0) = \vec{x}(init) + c_0\,\vec{\Delta}(0), \quad c_0 > 0, \tag{6.22}$$

for the next trial. Apply it to the process, acquire (observe) its output $y(0)$ and store it. Estimate an $N \times 1$ gradient vector $\vec{D}(0)$ with elements $D_i(0)$ as follows:

$$D_i(0) = \frac{y(0) - y(init)}{c_0 \Delta_i(0)}, \quad i = 1, 2, \ldots, N \tag{6.23}$$

Set pass number $k = 0$.

**Step 1** Compute the next decision sequence:

$$\vec{x}(k+1) = \vec{x}(k) - \alpha_k\,\vec{D}(k), \tag{6.24}$$

apply it to the process, then observe and store $y(k+1)$.

**Step 2** Generate the vector of random perturbations $\vec{\Delta}(k+1)$, form the trial decision sequence $\vec{x}(trial)$ as follows:

$$\vec{x}(trial) = \vec{x}(k+1) + c_k\,\vec{\Delta}(k+1), \tag{6.25}$$

apply it to the process, observe and keep its output $y(trial)$.

**Step 3** Estimate and store the gradient vector $\vec{D}(k+1)$ with elements that are computed as follows:

$$D_i(k+1) = \frac{y(trial) - y(k+1)}{c_{k+1}\Delta_i(k+1)}, \quad i = 1, 2, \ldots, N. \tag{6.26}$$

**Step 4** If $k < k_{max}$, set $k = k + 1$ and go to Step 1, otherwise: stop the algorithm and provide $\vec{x}(k)$ as the approximation of $\vec{x}^*$.

Several remarks are in order concerning the above algorithm.

- All the modifications and the generalizations of the Kiefer–Wolfowitz algorithm, including handling constraints, are applicable also to the SPADL (see the previous section).

- The SPADL fits the main guideline of this book, since the learning involves whole decision sequences when the transition from $\vec{x}_k$ to $\vec{x}_{k+1}$ takes place. There is, however, a subtle point, namely, each transition requires two runs (or simulations) of the process to be optimized in order to estimate the gradient at $\vec{x}_k$, instead of $2N$, as in the original Kiefer–Wolfowitz version.
- In this respect, the SPADL is very similar to the algorithm proposed in [168]. The distinction between them is in using the one-sided differences in the SPADL and the symmetric differences in [168] for estimating the gradient. As a result, we observe directly $y_k$ at $\vec{x}_k$, which is beneficial, but we may pay for this at a slightly slower rate of convergence of the SPADL. Nevertheless, the proof of convergence provided in [168] can be adapted to the SPADL, since the only difference appears in the upper bounds for the reminder of the Taylor expansion of $\mu(\vec{x})$.

We refer the reader to other variants of stochastic approximation algorithms in [25, 80, 101]. The latter paper deserves our attention, because it is based on a quite different paradigm than the main stream of research on stochastic approximation. Namely, instead of trying to estimate the gradient from as small number of trials as possible in [25] it is assumed that we have a large database of examples of a process behavior under different sequences of decisions, as it may happen, e.g., when an industrial installation is running for a long time. The authors propose to repeat many times drawing at random a subsets of the observations and to estimate the gradient by averages. This interesting new look at stochastic approximation is outside the scope of this book.

## 6.3 Response Surface Methodology for Searching for the Optimum

Response surface (RS) methodology emerged about seventy years ago. Its idea is close to the one discussed in this book, namely, a step-by-step improvement of a process with active learning between phases. Therefore, it is worth to pointing important features of the RS methodology. We refer the reader to [94, 95, 100] for its comprehensive description. The first applications of RS methodology took place in the chemical industry, while the later covered a large number other industrial and non-industrial processes.

The term *response surface* refers to the unknown function that links the response (the output, the yield or the quality criterion) of the process and its decision variables $\vec{x}$ (tunable parameters or inputs). The response $J(\vec{x})$ is observed at selected points $\vec{x}$'s with an additive random errors, which are uncorrelated, have zero mean and finite variances. The aim is to locate a minimum of the response surface by a sequence of experiments. Notice that this problem statement is very similar to t stochastic approximation in the Kiefer–Wolfowitz setting (see the previous subsection). However, there are subtle differences between these two approaches. Namely, RS methodology tacitly assumes that it is desirable to conduct carefully planned experiments between learning phases and there are suggestions how to select them. On the other hand, the number of between phases experiments is allowed to be larger than in simultaneous perturbation algorithms (see Sect. 6.2). Additionally, the phases of RS methodology are not only descent phases that are based on the gradient estimation, but also the quadratic approximation phase(s) for more precise location of the optimum.

## 6.3.1 Gradient Estimation According to Response Surface Methodology

To explain how the gradient of $J(\vec{x})$ at $\vec{x}_0$ is estimated from observations of $J(\vec{x})$ near $\vec{x}_0$ with random errors, consider the Taylor series of the first order, when $\vec{x}_0$ is perturbed by an $N \times 1$ vector $\overrightarrow{\delta x}$. This yields:

$$J(\vec{x}_0) + \overrightarrow{\delta x}) \approx J(\vec{x}_0) + \sum_{j=1}^{N} \frac{\partial J(\vec{x})}{\partial x_j}\bigg|_{\vec{x}=\vec{x}_0} \delta x_j, \tag{6.27}$$

where $x_j$ and $\delta x_j$, $j = 1, 2, \ldots N$ are elements of vectors $\vec{x}$ and $\overrightarrow{\delta x}$, respectively. Taking into account that $J(\vec{x})$ is observed with random errors, it is convenient to interpret (6.27) as a regression function:

$$y(\overrightarrow{\delta x}) = \beta_0 + \sum_{j=1}^{N} \beta_j\, \delta x_j, \tag{6.28}$$

where

$$\beta_0 \stackrel{def}{=} J(\vec{x}_0), \quad \beta_j \stackrel{def}{=} \frac{\partial J(\vec{x})}{\partial x_j}\bigg|_{\vec{x}=\vec{x}_0}, \quad j = 1, 2, \ldots, N. \tag{6.29}$$

Now, after collecting $y_i$ for $\overrightarrow{\delta x}_i$, $i = 1, 2, \ldots, n$, $\beta_j$'s are estimated by the classic least squares method (LSM), which provides their estimates $\hat{\beta}_j$'s. Thus, an $N \times 1$ vector $\hat{\beta}(0)$ with elements $\hat{\beta}_j$, $j = 1, 2, \ldots, N$ estimates the gradient at $\vec{x}_0$. Then, $\vec{x}_0$ is updated by making step in $-\hat{\beta}(0)$ direction and new $\overrightarrow{\delta x}$ is computed in a way to be described later (see [99]). Additionally, a statistical test (F-test is usually recommended) is applied to verify whether the linear model (6.28) is adequate. If not, it is inferred that we are near the optimum and the following quadratic model:

$$y(\overrightarrow{\delta x}) = \beta_0 + \sum_{j=1}^{N} \beta_j\, \delta x_j + \sum_{l,\, j=1,\, l \le j}^{N} h_{lj}\, \delta x_l\, \delta x_j \tag{6.30}$$

is fitted to observations by the LSM, after possible extensions of the number and positions of design points $\overrightarrow{\delta x}_i$'s. After applying the LSM, $\hat{\beta}(0)$ is interpreted as above, while estimates $\hat{h}_{lj}$'s of $h_{lj}$'s are used to approximate matrix $H(0)$ of the second derivatives of $y(\overrightarrow{\delta x})$ at $\vec{x}_0$. However, when the estimate $\hat{H}(0)$ of $H(0)$ is formed, one has to symmetrize it by setting $\hat{h}_{jl} = \hat{h}_{lj}$ for all $l > j$. Assuming that $\hat{H}(0)$ is positive definite (hence, also non-singular), one can obtain an approximation of the optimum by equating the gradient of

$$\overrightarrow{\delta x}^T\, \hat{\beta}(0) + \overrightarrow{\delta x}^T\, \hat{H}(0)\, \overrightarrow{\delta x} \tag{6.31}$$

to zero and solving the resulting linear equation with respect to $\overrightarrow{\delta x}$, which yields

$$\overrightarrow{\delta x}(0) = -\frac{1}{2}\, [\hat{H}(0)]^{-1}\, \hat{\beta}(0). \tag{6.32}$$

The phase index 0 in (6.32) would appear only if we were lucky to start in the close vicinity of the optimum solution. Otherwise, one can expect a larger index $k$. Computing $[\hat{H}(0)]^{-1}$ explicitly can be beneficial for diagnostic purposes.

### 6.3.2 RS Methodology—Learning Algorithm

An important ingredient of RS methodology is the selection of points $\overrightarrow{\delta x'}_i$, $i = 1, 2, \ldots, n'$ and $\overrightarrow{\delta x''}_i$, $i = 1, 2, \ldots, n''$, $n'' > n'$, at which the responses are measured for linear and quadratic approximation stages, respectively We shall call them Experiment Design 1 (ExpDes 1) and Experiment Design 2 (ExpDes 2), respectively. As routinely done in the literature on experiment design, it is further assumed that the designs are presented in the form normalized to the $N$-dimensional hypercube $\mathcal{C}_N = [-1,\ 1]^N$. After selecting experiment points in $\mathcal{C}_N$, they are moved to a proper design location, by translation of $\vec{0}$ to $\vec{x}_0$ in our case. Then, they are re-scaled, by linear transformations of $[-1,\ 1]$ to a minimum and maximum ranges of each variable, as measured in its natural units. In the simplest case we shall write $c_k\, \mathcal{C}_N$, if all the directions are re-scaled in the same way, i.e., to $[-c_k,\ c_k]$ interval. We assume that ExpDes1 and ExpDes2 are selected in such a way that all the parameters in (6.28) and (6.30) can be uniquely estimated by the LSM. Further discussion on selecting these designs is provided after presenting an algorithm of learning extremum by the RS methodology (RSM).

**RSM-Based Algorithm**

**Preparations:** Select initial guess $\vec{x}_0$ and sequences $\alpha_k$' and $c_k$'s for which conditions (6.5) hold. Choose experiment designs ExpDes1 and ExpDes2. Select also a stopping condition, in the simplest case by bounding the number of updates by $k_{max} > 1$.

**Initialization:** Move ExpDes1 to decision sequence $\vec{x}_0$. Apply input sequences $\vec{x}_0 + c_0\,\overrightarrow{\delta x'}_i$, $i = 1, 2, \ldots, n'$ to the process and acquire (observe) its outputs (responses) $y_{0i}$'s and store them. Apply the LSM to these data so as to obtain the estimate $\vec{\beta}(0)$ of the gradient at $\vec{x}_0$. Set pass number $k = 0$.

**Step 1** Compute the next decision sequence:

$$\vec{x}(k+1) = \vec{x}(k) - \alpha_k\,\vec{\beta}(k), \tag{6.33}$$

as the center for the next series of experiments.

**Step 2** Move ExpDes1 to $\vec{x}(k+1)$. Apply input sequences $\vec{x}(k+1) + c_k\,\overrightarrow{\delta x'}_i, i = 1, 2, \ldots, n'$ to the process and acquire (observe) its outputs (responses) $y_i(k+1), i = 1, 2, \ldots, n'$ and store them.

**Step 3** Apply the LSM to these data so as to obtain the estimate $\vec{\beta}(k+1)$ of the gradient at $\vec{x}(k+1)$. Use test F to check whether the linear model is still adequate. If it is, go to Step 4, otherwise, go to Step 5.

**Step 4** If $k < k_{max}$, set $k = k + 1$ and go to Step 1, otherwise: stop the algorithm and provide $\vec{x}(k+1)$ as the approximation of $\vec{x}^*$.

**Step 5** If $k \geq k_{max}$ stop the algorithm and provide $\vec{x}(k+1)$ as the approximation of $\vec{x}^*$. Otherwise, move ExpDes2 to $\vec{x}(k+1)$ and apply input sequences[4] $\vec{x}(k+1) + c_k\,\overrightarrow{\delta x''}_i, i = 1, 2, \ldots, n''$ to the process and acquire (observe) its outputs (responses) $y_i(k+1), i = 1, 2, \ldots, n'$ and store them. Fit the second order model (6.30), applying the LSM to these data. As result, obtain new[5] estimate$\widehat{\vec{\beta}}(k+1)$ and $\hat{H}(k+1)$. If this matrix is nonsingular, update $\vec{x}(k+1)$ as follows:

$$\vec{x}(k+1) - \frac{1}{2}\,[\hat{H}(k+1)]^{-1}\widehat{\vec{\beta}}(k+1) \tag{6.34}$$

and stop the algorithm, providing (6.34) as the result. Optionally, one can repeat Step 5 from the beginning, replacing $\vec{x}(k+1)$ by (6.34).

As already mentioned, sequences $\alpha_k$' and $c_k$'s for which conditions (6.5) hold. The reader is referred to Fig. 6.1 and to [144], in which the choice of $\alpha_k = \varsigma_1/k$, $\varsigma_1 > 0$ is recommended.

Possible repetitions of Step 5 are similar to the sequential quadratic programming (SQP) algorithm (see [105, 106] for its comprehensive description and [131] for the SQP algorithm with a modified Fletcher's filter).

---

[4] If ExpDes2 is a composite experiment design (see [99]) that is based on ExpDes1, then one can save a part of inputs applied at this step.

[5] Notice that$\widehat{\vec{\beta}}(k+1)$ differs from $\vec{\beta}(k+1)$, since they are estimated from observations obtained from different designs ExpDes2 and ExpDes1, respectively. Furthermore, $\widehat{\vec{\beta}}(k+1)$ is estimated together with $H$ matrix.

All the modifications and constraints handling approaches, including the population filter, that are described in Sects. 6.1.3 and 6.1.4 are applicable also to the above algorithm.

### 6.3.3 Selecting Experiment Designs

The starting point for the selection of experiment design ExpDes1 for the first stage is the full factorial design at two levels, denoted as $2^N$ design. As is known (see e.g. [94]), entries $\overrightarrow{\delta x'}_i$, $i = 1, 2, \ldots, n' = 2^N$ of the $2^N$ design consists of all the combinations of $\pm 1$ . This design has a desirable property, namely, it is orthogonal in the sense that for design matrix $X'$ the corresponding Fisher information matrix (FIM), normalized by the variance of observation errors, fulfills:

$$X'\,(X')^T = 2^N\, I_N, \quad \text{where } X' \stackrel{def}{=} \left[\overrightarrow{\delta x'}_1,\, \overrightarrow{\delta x'}_2, \ldots, \overrightarrow{\delta x'}_{2^N}\right]. \tag{6.35}$$

This design is also D-optimal, which means that it has the largest determinant of $X'\,(X')^T$ matrix among all $2^N$ point designs with entries in the $[-1,\ 1]^N$ hypercube. Nevertheless, from our point of view, these desirable properties of $2^N$ designs are highly overshadowed by a huge number of experimental runs of $2^N$ full factorial designs (over one million for as short a decision sequence as $N = 20$). This is in sharp contrast with the number of $(N + 1)$ parameters to be estimated in the linear model. For this reason, one may try to apply fractional factorial designs at two levels that are obtained from $2^N$ design by selecting a fraction (usually one half, one fourth etc.) of columns from $X'$ in such a way that they are still orthogonal and simultaneously, they allow for estimating $\beta_0, \beta_1, \ldots, \beta_N$ without confounding them with parameters corresponding to interaction terms.

Another way of conducting orthogonal experiments for linear models is called the simplex design that is concentrated at $(N + 1)$ vertices of the $N$-dimensional simplex, which is centered at the origin (see [94] Sect. 11.4.1). This design is saturated, which means that it has the same number of design points as the number of estimated parameters.

The choice of ExpDes2 for estimating all the parameters in quadratic model (6.30) is even more demanding, since one has to estimate $1 + N + N\,(N - 1)/2$ parameters with at least the same number of experimental runs of the process. Observe that two level $\pm 1$ designs are not appropriate for this purpose, since the quadratic terms are always at the level of 1, precluding the possibility of estimating parameters corresponding to them.

Formally, full factorial designs at three levels $\{-1, 0, 1\}$, denoted as $3^N$, would be appropriate, but the gap between the number of runs and the number of parameters to be estimated grows even faster than for $2^N$ designs. As the result, $3^N$ designs can be useful for very short decision sequences of the length $N = 3$ or $N = 4$, except for special cases when running an optimized process is cheap and not too time-consuming.

The well-known remedy is the use of the so-called central composite designs (see e.g. [94, 99]). These designs are composed by adding sets of three kinds, namely, the set of vertices of the $2^N$ design or a selected fractional design, the repetitions at the origin and the set of the so-called *star points* that are placed along each axis as follows:

$$[\pm \upsilon, 0, \ldots, 0]^T, [0, \pm \upsilon, \ldots, 0]^T, \ldots [0, 0, \ldots, \pm \upsilon]^T, \tag{6.36}$$

where $\upsilon > 0$ is selected by the experimenter, frequently in such way that all the composite design has the property known as a *rotatability*. This means that variance $\mathbb{V}ar(\vec{x})$ of estimating the quadratic function is the same for all the points on each sphere centered at the origin. In other words, $\mathbb{V}ar(\vec{x})$ is a function that depends on $\vec{x})$ only through $||\vec{x})||$. It can be proved that $\upsilon = \sqrt{n_f}$ provides the rotatable design.

Central composite designs are recommended as those with a reasonable number of additional experiments. Notice that they nicely fit in the Algorithm of learning extremum by RS methodology, since when passing from Step 4 to Step 5 we already have the fractional factorial part of the composite design.

Another class of experiments that are parsimonious in using them is known as Box–Behnken designs—see [94, 99] for their properties.

Discussing experiment designs that can be used for optimizing decisions, it is worth mentioning the approach developed in [116–118]. The authors called them *partition experimental designs for sequential processes* for the first and the second order regression models as well as for a process with a large number of variables, respectively. In these papers, the term *sequential processes* refers to serially connected blocks that form the process as a whole. An appealing feature of the proposed partition experiment designs is that the authors consider one experiment design with good properties for the whole process. Then this design is partitioned in such a way that their columns correspond to inputs applied to particular blocks. This is possible because each block has its own set of input variables that are non-overlapping.

The problem setting considered in this book can also be depicted as serial connections of blocks that represent subsequent passes of a repeatable process. There is, however, a fundamental difference between this setting and the one mentioned above, namely, in our setting all the virtual blocks have the same set of inputs and only their values change from pass to pass. Thus, we cannot use directly the results proposed in [116–118]. They can serve, however, as methodological guidelines.

## 6.4 Discussion on Stochastic Gradient Descent Approaches

The algorithms:

**K-WSAA** —the Kiefer–Wolfowitz stochastic approximation algorithm,
**SPADL** —the simultaneous perturbation algorithm for decision learning,
**RSM** —the algorithm of learning extremum byRS methodology

have the same general structure of iteration-to-iteration learning that is sketched as a flowchart in Fig. 6.2. Their common features are the following:

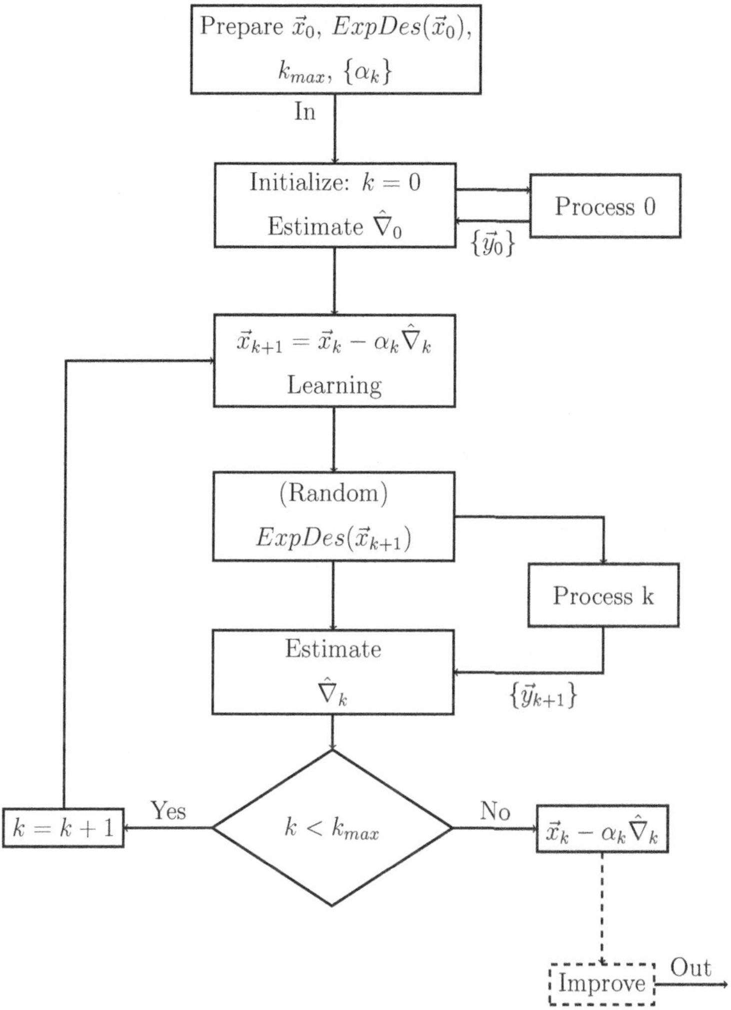


**Fig. 6.2** A general flowchart of stochastic gradient algorithms

- the main loop of learning using the stochastic gradient descent,
- the necessity of interactions(information exchange) between the algorithm and the process to be optimized.

There are, however, important differences in details. Namely, the number of passes of the process needed for estimating the gradient at a given point:

- only one for the SPADL,
- $N$ when the K-WSAA is used (or 2 $N$ in the original symmetric difference version),
- at least $(N + 1)$, if a saturated is experiment design is selected in the RSM-based algorithm.

The second difference is in the statistical accuracy of estimating the gradient, as measured by the determinant of the FIM. Here, the hierarchy of these algorithms is quite the reverse, since the RSM-based algorithm, even if not using the D-optimal design, is usually optimized in this direction, the designs used in the K-WSAA are known to be less efficient from this point of view. Probably, this fact motivated Kiefer and Wolfowitz to develop an elegant theory of optimal experiment designs. Finally, randomized designs used in the SPADL have D-efficiencies fluctuating from pass to pass. This feature can be considered either as a drawback or as a chance to locate a global optimum.

The notation $\hat{\nabla}_k$ in Fig. 6.2 is used to denote the estimate of the a goal function gradient that is obtained by one of the above listed methods at $k$th iteration (pass).

### 6.4.1 Selecting the Step Length and Scaling

An important component of stochastic gradient descent algorithms is the selection of step length $\alpha_k$ or its component-wise counterparts. In addition to methods already mentioned, a large stream of research is devoted to the so called adaptive choice of $\alpha_k$'s, which—roughly speaking—means that $\alpha_k$ is selected in a way depending on observations of the goal function and its derivatives. Notice that most of methods developed in the optimization algorithms community are based on a line search and/or evaluated hessian inverse. Their applicability for our purposes is limited, since they usually require additional passes (runs) of the optimized process that is rarely acceptable.

In recent ten years several new methods of selecting the step length and/or scaling the gradient components have been developed researchers from deep neural network laboratories and they are at present an active area of investigations. Furthermore, they are of vital interest for learning decision sequences, since they are solely based on $\hat{\nabla}_k$ and its previous estimates, without the need for additional runs of the optimized process.

The first of these methods was the adaptive gradient algorithm [39]. Its acronym ADAGRAD is in a wide use. The idea of ADAGRAD is to scale each of the gradient component $\hat{\nabla}_k^{(n)}$ by

$$w_k^{(n)} \stackrel{def}{=} \sum_{j=1}^{k} [\hat{\nabla}_j^{(n)}]^2, \quad n = 1, 2, \ldots, N, \tag{6.37}$$

which leads to the following learning algorithm

$$\vec{x}_{k+1}^{(n)} = \vec{x}_k^{(n)} - \frac{\alpha}{\sqrt{w_k^{(n)}}} \hat{\nabla}_k^{(n)}, \quad n = 1, 2, \ldots, N, \tag{6.38}$$

where $\alpha > 0$ is a preselected constant.

The second algorithm in this group is known as the Root Mean Square Propagation (RMSProp). It is based on the EWMA averaging of $[\hat{\nabla}_j^{(n)}]^2$'s that are inserted into (6.38) instead of $w_k^{(n)}$.

A large group of algorithms, known under the name Adam, NAdam etc. [187], was obtained by applying linear combinations of $\hat{\nabla}_j^{(n)}$'s in the nominator of the updating term and $[\hat{\nabla}_j^{(n)}]^2$'s in its denominator.

In practice, all these algorithm are known to be very efficient, even for non-convex goal functions [180] and one can expect that they be leading methods for large stochastic gradient optimization problems. On the other hand, a care should be kept, since it is also known that even for convex goal functions a lack of convergence can happen [143].

# Chapter 7
# Iterative Learning of Optimal Decision Sequences

This chapter is—in some sense—central to this monograph, since it propounds and illustrates the notion of the iterative learning of optimal decision sequences (ILODS). Earlier, in Sects. 7.1 and 7.2, brief reviews of the run-to-run and iterative learning control (ILC) streams of research are presented. They serve as motivations and inspirations for the ILODS algorithm proposed in Sect. 7.3.

On the other hand, the main difference between the ILODS approach and the algorithms proposed in the previous chapters is that it uses explicitly a model of process dynamics. As a result, the ILDOS algorithm is able to learn much longer decision sequences. However, it is worth mentioning that one can also apply differential evolution and stochastic gradient algorithms to dynamical systems, but their dynamics are "hidden" in processes and only a decision sequence and its influence on the final result can be observed. In such circumstances, our ability to optimize longer decision sequences is largely reduced.

## 7.1 Run-to-run Control as an Inspiration

A common feature of the run-to-run control and the ILC approaches is their ability to learn from pass to pass (or between passes) of the process. To this end, when the present pass is finished, information on its quality and behavior is measured and transmitted to an optimization unit that transforms it into improvements of a decision sequence to be applied at the next pass. This feature does not exclude possible applications of actions along each pass, as done in classic control systems, by using e.g., PID controllers. However, for clarity of the presentation, this aspect is no longer discussed in this chapter.

The ability of learning between passes is also built into the repetitive control systems. Their aim is to learn periodic disturbances in order to suppress them. We



W. Rafajłowicz, *Learning Decision Sequences For Repetitive Processes—Selected Algorithms*, Studies in Systems, Decision and Control 401,
https://doi.org/10.1007/978-3-030-88396-6_7

shall not discuss them in the context of the run-to-run process control, since their goals are similar.

### 7.1.1 Learning in Run-to-run Decision Systems

The idea of run-to-run control arose at the beginning of the 1990s [153] and was applied in the semiconductor production industry [98]. The process of producing semiconductor wafers needs subtle tuning, but one has very limited possibilities to observe its partial results. Therefore, the tuning of the process parameters can be done based on measurements of the final result of a given run only. Similar circumstances appear also in many batch processes of chemical engineering. There, one can influence the temperature of reactants and the intensity of stirring them at subsequent reaction stages, but—again—the result can be evaluated only after finishing a given batch. For this reason, attempts to improve such processes are called batch-to-batch control. It seems, however, that the idea of run-to-run decision improvements can have much wider applications than tuning production processes, including health recovery processes and the growth of the economy, considered from month-to-month or year-to-year. We refer the reader to survey papers [87, 179] for examples of other applications, the variety of mathematical models and approaches to learning according to the run-to-run principle.

Traditionally, relatively simple linear regression models are used to explain the run-to-run approach (see, e.g. [179]). Namely, at $k$th run (pass) output (reaction) $\vec{y}(k)$ of a process is related to a sequence of decisions (manipulated parameters) $\vec{x}(k)$ as follows:

$$\vec{y}(k) = C\,\vec{x}(k) + \vec{b}(k) + \vec{\epsilon}(k), \quad k = 1, 2, \ldots \tag{7.1}$$

where $C$ is a matrix of constants that is assumed to be known, either from theoretical considerations or from the estimation based on pre-runs, $\vec{\epsilon}(k)$ is a vector of random disturbances with zero means and finite variances, while $\vec{b}(k)$ is a vector of unknown or uncertain parameters. This vector may represent the model uncertainty, a slowly varying drift or, an aging process and, therefore, there is a need for learning $\vec{b}(k)$ from previous runs at our disposal, i.e., only matrix $C$ is available, while pairs:

$$(\vec{x}(k),\ \vec{y}(k)), \quad k = 1, 2, \ldots \tag{7.2}$$

arrive to a learning system subsequently, after finishing each run.

### 7.1.2 Outline of the Run-to-run Optimization Algorithm

The most frequently recommended learning algorithm is the exponentially weighted moving average (EWMA) of the following form: for $k = 1, 2, \ldots$ iterate

$$\hat{\vec{b}}(k) = (1-\gamma)\,\hat{\vec{b}}(k-1) + \gamma\left[\vec{y}(k) - C\,\vec{x}(k)\right], \tag{7.3}$$

where

$\hat{\vec{b}}(0)$ is our initial guess for $\vec{b}$,
$\hat{\vec{b}}(k)$ is the estimate of $\vec{b}(k)$ after $k$th run (pass of the process),
$0 < \gamma < 1$ is the tuning parameter that dictates the rate of forgetting previous estimates of $\vec{b}$,
$\vec{y}(k) - C\,\vec{x}(k)$ is the current (rough) estimate of $\vec{b}(k-1)$, motivated by (7.1) with neglected $\vec{\epsilon}(k-1)$ term.

As is known, the EWMA scheme, when the learning process is convergent, is able to recover a constant vector. Furthermore, it acts as a low pass filter with the cut-off frequency dependent on $0 < \gamma < 1$ and the period between samples (between subsequent runs in our case, see [132] for more details on the interpretation of the EWMA as a filter).

Let $\vec{y}^*$ be a given target vector for the process output. Consider the simplest case when the dimension of $vecy(k)$ is the same as the dimension of $\vec{x}(k)$. Then, $C$ is a quadratic $N \times N$ matrix that is further assumed to be nonsingular. To motivate further considerations, assume for a while that $\vec{b}$ is known and there are no random errors. In such a case we obtain:

$$\vec{y}^* = C\,\vec{x} + \vec{b} \;\Rightarrow\; \vec{x}^* = C^{-1}\,(\vec{y}^* - \vec{b}) \tag{7.4}$$

and the proper decision sequence $\vec{x}^*$ would be immediately known. According to the plug-in idea, from the second equality in (7.4) we obtain

$$\vec{x}(k) = C^{-1}\,(\vec{y}^* - \hat{\vec{b}}(k-1)), \tag{7.5}$$

where $\vec{x}(k)$ is an estimate of $\vec{x}^*$ at $k$th run.

Summarizing, the run-to-run updating of the decision sequence consists of the estimation (the EWMA filtering) part (7.3) and the updating (7.5). Again, each run requires one pass of the process and collecting measurements from its output—see Fig. 7.1 a general flowchart.

This flowchart also covers more complicated cases when $C$ is not an $N \times N$ matrix. We refer the reader to [98] pp 72–74 for an analysis of cases when the dimension of $\vec{y}(k)$ is either larger or smaller than $N$ (see also [179]). In [62] the important case of univariate responses $y(k)$'s is considered (in this case $C$ is a vector). Also sufficient conditions for the convergence of the learning process are provided and discussed. In most cases, run-to-run decisions optimization is, intentionally, considered for simple and static models, such as that described by (7.1), tacitly assuming that possible transient phenomenons vanish during the run. More recently, see [34] and the bibliography therein, models of dynamic processes are used for optimizing decisions. Such approaches are similar to the iterative learning control (ILC) ideas that are discussed in the next section. Similarities between run-to-run and the ILC algorithms have already been revealed (see [179].

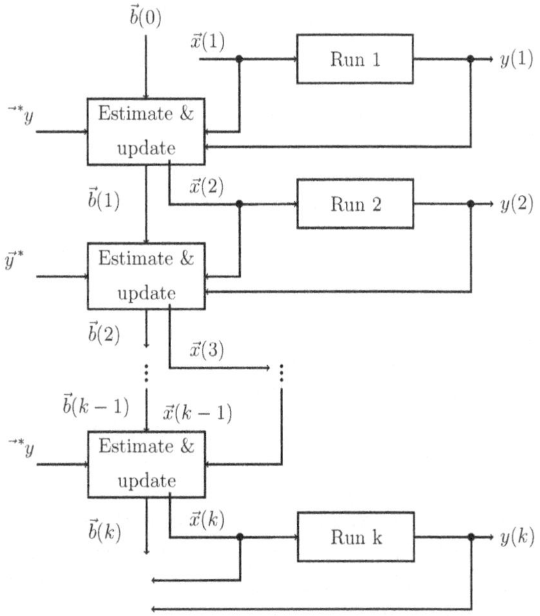


**Fig. 7.1** A general flowchart of run-to-run decisions updating

## 7.2 Iterative Learning Control—In Brief

The next important research stream that develops the idea of learning from pass to pass is iterative learning control. It emerged in control systems theory in the 1970s, but wide research started in the middle of the 1980s and they are still intensively developed. We refer the reader to [111] for the history, basic notions and the survey of the optimization approach to the ILC.

It seems that the main difference between the run-to-run optimization and the ILC approach is in assumptions concerning:

1. a model of a process, which is a dynamic one in ILC theory and a static one in most cases of the former,
2. observations that are available during the process run, namely, they usually include measurements of the process state at each time instant along each pass, while in the run-to-run optimization frequently only the overall quality criterion of each run is measurable,
3. the possibilities of influencing the process behavior along the pass are frequently much larger in ILC theory.

An important similarity of both approaches is in the method of formulating the goal of learning. Namely, in the classic formulations of them, the goal is to learn the sequence of inputs in such a way that the sequence of outputs (responses) is convergent to that desired. In the ILC theory, this desired sequence is called the reference signal.

### 7.2.1 Basic Formulation of the ILC

To sketch basic notions of the ILC approach, consider the following simple model of a repetitive process at pass $k = 0, 1, \ldots$ and discrete time $n = 0, 1, \ldots, (N-1)$

$$s_{n+1}(k) = A\, s_n(k) + b\, a_n(k), \quad s_0(k) = \vec{0}, \tag{7.6}$$

$$y_n(k) = c^T s_n(k), \tag{7.7}$$

where, for the simplicity of the exposition, sequences of decisions (actions) $a_n(k)$ and responses (outputs) $y_n(k)$ are univariate, $n = 0, 1, \ldots, (N-1)$. The remaining symbols in (7.6) are standard ones, namely,
– process state $s_n(k)$, at pass $k$ and time $n$ along the pass, is a $d_s \times 1$ vector of real valued variables,
– $A$ is a $d_s \times d_s$ transition matrix,
– $b$ is a column vector $d_s \times 1$ of actions amplifications,
– $c$ is a column vector $d_s \times 1$ that indicates how states influence the output.

Let us assume that a desired behavior of process (7.6), (7.7) is given as the reference sequence $r_n$, $n = 1, 2, \ldots, N$ for its output. Notice that this sequence does not depend on the pass number.

In the simplest version, the problem of the ILC is to derive a learning algorithm that updates sequences of decisions

$$\vec{a}(k) = [a_0(k), a_1(k), \ldots, a_{(N-1)}(k)]^T \tag{7.8}$$

in such a way that sequence of error vectors $\vec{e}(k)$, defined as follows:

$$\vec{e}(k) = [r_1 - y_1(k), r_2 - y_2(k), \ldots, r_N - y_N(k)]^T \tag{7.9}$$

is convergent to zero as $k \to \infty$. In other words, it is expected that $\lim_{k\to\infty} ||\vec{e}(k)|| = 0$, i.e., the algorithm asymptotically learns how to attain a desired trajectory of the output. This formulation tacitly assumes that there exists sequence of decisions $\vec{a}^*$ and the corresponding sequence of states $\{s_n^*\}$, generated by process (7.6) such that

$$r_n = c^T s_n^*, \quad n = 1, 2, \ldots, N. \tag{7.10}$$

The conditions for the existence of such $\vec{a}^*$ are well known as the full output controllability and we omit the details here, assuming that they are fulfilled. Hence, it is reasonable to require that the learning algorithm assures also

$$\lim_{k\to\infty} ||\vec{a}^* - \vec{a}(k)|| = 0 \tag{7.11}$$

as usually required in ILC theory.

Several remarks are in order concerning (7.6).

- In (7.6) zero initial conditions are assumed without losing the generality. Indeed, due to the linearity of model (7.6), (7.7), (7.9), if nonzero initial conditions $s_0(k) = s_0$ are met, it suffices to incorporate them into the reference sequence, substituting $r_n - c^T A^n s_0$ instead of $r_n$, $n = 1, 2, \ldots, N$.
- Observe that (7.6) is the so-called 2D system in the sense that state $s_n(k)$ is a function of two independent variables, namely, $k$ and $n$. The theory of 2D systems plays an important role in the development of the ILC (see e.g., [56, 57, 112, 114]). Also its generalization to nD systems was used for the ILC of spatio-temporal processes [32, 33], but we shall not go into detail in this book.
- The condition

$$\lim_{k\to\infty} ||\vec{e}(k)|| = 0 \tag{7.12}$$

  is a relatively weak one. It does not impose any requirements on the behavior of $e_n(k) = r_n - y_n(k)$ along the pass, i.e., considered as the sequence of $n = 1, 2, \ldots, N$ for fixed $k$, except that $||\vec{e}(k)|| < \infty$. We refer the reader to [147] for deep results on related topics.
- Model (7.6) does not take into account the possible influence of earlier process passes on the current pass. The reader is again referred to [147] for such models.

When $k$th pass is finished, we have at our disposal $\vec{e}(k)$, $\vec{a}(k)$ and their previous copies. Thus, they can be used for forming $\vec{a}(k+1)$. There is, however, a subtle point that is worthy of attention, since it is crucial for understanding the strength of the ILC approach. It can be illustrated by the classic learning rule, known as the Arimoto algorithm:

$$a_n(k+1) = a_n(k) + \gamma\, e_{n+1}(k), \quad n = 0, \ldots, (N-1), \tag{7.13}$$

where $\gamma$ is an amplification coefficient. The crucial point is that (7.13) looks like a predictive algorithm, since at time instant $n$ the correction of $a_n(k+1)$ depends on the error at time instant $(n+1)$. However, this algorithm is a non-anticipating one, since $e_{n+1}(k)$ is already known when $k$th pass is finished. On the other hand, if disturbances are small, one can expect that this quasi-predictive feature will be beneficial and it is, since for properly selected $\gamma$ (see below) this learning algorithm is convergent. Nevertheless, this convergence is not monotone.

**Remark 7.9** It has been known for a long time that at the beginning of learning $||\vec{e}(k)||$ may largely grow and then drops toward zero. Many efforts of researchers have been undertaken to reduce or preclude this, undesirable in practice, effect. We shall mention them later.

The Arimoto rule is a special case of the so-called P-type ILC algorithms of the following form:

$$\vec{a}(k+1) = \vec{a}(k) + \beta\, K\, \vec{e}(k), \tag{7.14}$$

where $K$ is an $N \times N$ matrix of weights, while $\beta$ is an amplification factor. In particular, (7.13) falls into this class, if $K$ is chosen to an $N \times N$ zero matrix, except one band above the main diagonal, which is filled with $\gamma$ as entries.

From the formal point of view, learning algorithm (7.14) is similar to those discussed in the previous chapter. In practice, however, there is an important difference in the size of $N$, which is frequently larger than thousands, e.g., when a process states are sampled with the frequency of 1 kHz and a pass length has several seconds. Additionally, as a consequence, the length of the decision sequence may have the same length.

The classic tool for investigating the convergence of the ILC is so-called lifting (see e.g., [147]). Its essence is in solving the solution of (7.6) and (7.7), with respect to $n$, explicitly and in writing down the iterates with respect to $k$ using so-called super-vectors. In the single input, single output case it suffices to use (7.8). Let us also assume that (7.6) and (7.7) is of the relative degree one. Then see e.g., [31]

$$\vec{y}(k) = G\,\vec{a}(k), \quad \text{and} \quad \vec{e}(k) = G\,(\vec{a}^{*} - \vec{a}(k)), \tag{7.15}$$

where an $N \times N$ matrix $G$ is lower triangular and has the following entries

$$G = \begin{bmatrix} c^T b & 0 & 0 & \dots & 0 \\ & c^T b & 0 & \dots & 0 \\ & & \ddots & \dots & 0 \\ & \mathbf{c^T A^j b} & & c^T b & 0 \\ & & & & c^T b \end{bmatrix}, \quad j = 1, 2, \dots, \tag{7.16}$$

where $\mathbf{c^T A^j b}$ stands for the typical element of the columns below the main diagonal, starting from $j = 1, 2, \dots, (N-1)$ and finishing with $j = 1$ in the column before the last one. Representation (7.15) is valid also when the relative degree is larger than one (see [1, 113]), assuming that matrix $G$ is appropriately modified (see formula (15) in [113]).

Multiplying (from the left) both sides of (7.14) by $G$ and using (7.15), we immediately obtain:

$$\vec{y}(k+1) = \vec{y}(k) + \beta\, G\, K\, \vec{e}(k), \tag{7.17}$$

which yields the pass-to-pass relationship between errors:

$$\vec{e}(k+1) = [I_N - \beta\, G\, K]\,\vec{e}(k), \quad k = 1, 2 \dots. \tag{7.18}$$

Assuming that $\beta$ and $K$ are selected in such a way that the spectral radius of matrix $[I_N - \beta\, G\, K]$ is less than one (strictly), we infer that the sequence of errors is asymptotically convergent. However, for large $N$ this condition is difficult to verify, since it requires calculating the eigenvalue of $[I_N - \beta\, G\, K]$ with the largest absolute value. It can be useful when one is able to calculate the eigenvalues analytically, as is the case of the Arimoto rule (set $\beta = 1$, $K = I_N$, for which the spectral condition yields that $\gamma$ should be selected in such a way that $|1 - \gamma\, c^T\, b| < 1$.

The second drawback of the spectral radius condition is that it may happen that $||\vec{e}(k+1)|| > ||\vec{e}(k)||$ for a finite number of positive integers $k$. Imposing a slightly more restrictive condition that for a certain matrix norm, induced by a vector norm in $R^N$, we have:

$$||[I_N - \beta\, G\, K]|| \leq q < 1, \tag{7.19}$$

we can characterize the sequence $||\vec{e}(k)||$'s in more details. Indeed, from (7.18) and (7.19) we obtain: $||\vec{e}(k+1)|| \leq q\, ||\vec{e}(k)||$. Thus $\lim_{k\to\infty} ||\vec{e}(k)|| = 0$ and the convergence is monotone. Furthermore, by iterating $||\vec{e}(k+1)|| \leq q\, ||\vec{e}(k)||$ for the rate of convergence we have:

$$||\vec{e}(k)|| \leq q^k\, ||\vec{e}(0)||, \quad k \geq 1. \tag{7.20}$$

For $c^T\, b \neq 0$ one can also infer the convergence of learning decision sequence, since $G$ is invertible[1] and for the induced matrix norm we obtain from the second equality in (7.15)

$$||(\vec{a}^* - \vec{a}(k))|| = ||G^{-1}\,\vec{e}(k)|| \leq ||G^{-1}||\, ||\vec{e}(k)||, \quad k = 1, 2, \ldots, \tag{7.21}$$

which implies $\vec{a}(k) \to \vec{a}^*$ as $k \to \infty$ with the same rate as in (7.20). This, in turn, implies also the convergence of process outputs (reactions) $\vec{y}(k)$'s to $\vec{y}^*$.

It remains to point out how to verify condition (7.19). Fortunately, the matrix norms induced by the max and the sum of absolute values vector norms are directly expressible in terms of absolute values of matrix elements. The Euclidean norm induces the matrix norm that is upper bounded by the Frobenius matrix norm, which is again directly expressible by matrix elements. Thus, even for large $N$ one is able to check condition (7.19) for given $\beta$ and $K$. However, if this condition does not hold, we have no indication how to correct $\beta$ and $K$. This was probably one of the reasons why optimization-based approaches were developed.

In the class of linear ILC schemes more advanced versions of (7.14) are considered such that $\vec{a}(k+1)$ is a linear combination of several previous $\vec{a}(k)$'s and $\vec{e}(k)$'s.

We refer the reader to [113] for more advanced learning algorithms of the gradient type and conditions of their convergence that take into account robustness against model inaccuracies. This is one of the main research topics in ILC theory, since it is—in most cases—a model-based approach. Linear matrix inequalities (LMI) is a powerful tool used for designing robust control algorithms (see e.g., [169]).

It is worth noticing that (7.14) is a feedforward algorithm in the sense that it uses only data from past sequences of errors. In ILC theory also combined feedforward/feedback rules are considered (see [148] and the bibliography cited therein). Such rules apply not only to data from past passes, but also observations of errors from the current pass that is fed back in order to reduce current, non-repeatable disturbances. We shall not follow such rules in the sequel, since our focus is on

[1] We shall use $G^{-1}$ for theoretical purposes only, since computing the inversion of $G$ can be a tremendous task.

pass-to-pass learning of whole decision sequences. This statement does not exclude feedback type realizations of decision sequences in the following way. After updating $\vec{a}(k+1)$, simulate the expected process response $\vec{y}(k+1)$ and provide it as a temporary (for pass $(k+1)$) reference signal to a feedback controller, whose aim is to generate a decision sequence that is supplied to the process.

### *7.2.2 An Optimization Paradigm in ILC Theory*

The optimization paradigm was introduced to ILC theory (see [110, 111] and the bibliography cited therein) as a remedy for the drawbacks of a heuristic design of ILC systems that are mentioned in Remark 7.9. The idea is to introduce the sequence of learning quality criterions of the from: for $k = 1, 2, \ldots$

$$J_{k+1}(\vec{a}(k+1)) = ||\vec{e}(k+1)||^2 + ||\vec{a}(k+1) - \vec{a}(k)||^2, \tag{7.22}$$

where $||.||$ stands for the Euclidean norm[2] and

$$\vec{e}(k+1) = \vec{r} - G\,\vec{a}(k+1), \quad \vec{r} = [r_1, r_2, \ldots r_N]^T. \tag{7.23}$$

The minimization of (7.22) with respect to $\vec{a}(k+1)$, taking into account (7.23) and possible constraints imposed on $\vec{a}(k+1)$ is known as the norm optimal ILC (NOILC). The minimizer is further denoted as $\vec{a}^\star(k+1)$. Notice that the minimization of (7.22) is performed after each pass. Its aim is to balance the rate of convergence of $||\vec{e}(k)||$ to zero and to prevent too large changes of $\vec{a}(k)$ between subsequent passes.

Under weak assumptions imposed on $G$, it can be shown (see [111] and the bibliography cited therein) $\vec{a}^\star(k+1)$'s ensure monotonic convergence of $||\vec{e}(k)||$ to zero with the geometric rate. Furthermore, these results convey to abstract Hilbert spaces [110].

Among many useful approaches to the NOILC problem we mention [6, 156] that are of importance from our point of view. The reason is the idea of computing the gradient of the NOILC objective function (7.22) by applying the co-state (adjoint) equations. As demonstrated in [6, 156], this approach leads to efficient algorithms of improving long decision sequences.

## 7.3 Iterative Learning of Optimal Decision Sequences

Objective function (7.22), discussed in Sect. 7.2.2, is selected from the point of view of the proper behavior of errors in pass-to-pass learning. However, it is worth

---

[2] In (7.22) one can use more general norms: $(\vec{z}\,R\,\vec{z})^{1/2}$, $(\vec{z}\,S\,\vec{z})^{1/2}$ with positive definite matrices $R$ and $S$.

considering the optimization paradigm for more general objective functions $J(\vec{a})$, assuming that this function is continuously differentiable and strictly convex in an $N$-dimensional Euclidean space. The dependence of $J$ on $\vec{a}$ can be both a direct one and an indirect one, by state equations. Here, we shall write state equations in the form that frequently arises when process originally described by the ODE's is approximated by finite differences in a simple way. Namely, for $n = 0, 1, \ldots, (N-1)$,

$$\vec{s}_{n+1} - \vec{s}_n = A\,\vec{s}_n + b\,a_n, \quad \vec{s}_0 = \vec{0}, \tag{7.24}$$

where—for simplicity of formulas—the linear system is selected, with $A$ and $b$ having the same description as in Sect. 7.2.1. Also the objective function is selected in the standard form:

$$J(\vec{a}) = \sum_{n=0}^{N-1} \phi(\vec{s}_{n+1}, a_n), \tag{7.25}$$

where $\phi : R^{d_s} \times R \to R^+$ takes nonnegative values and it evaluates the quality of current decision $a_n$ and the next state $\vec{s}_{n+1}$, resulting from this decision. We assume that $\phi$ is continuously differentiable w.r.t. both arguments.

A standard formulation of the optimal control problem is the following: find sequence $\vec{a}^* = [a_0^*, a_1^*, \ldots, a_{N-1}^*]^T$ that minimizes $J(\vec{a})$ under constraints (7.24) and possible other constraints imposed on $a_n$'s and/or on $\vec{s}_n$'s that are assumed to be already incorporated into $J$ as penalties—again for simplicity of formulations.

There is a huge bibliography of papers and monographs on solving numerically this and even much more advanced problems of optimal control for discrete time systems (see e.g., [17, 18, 124, 148] and the bibliographies cited therein). From the variety of proposed approaches we select one that is based on computing the gradient of the Lagrange function (or the Frechet derivative when processes with continuous time are considered).

Define the Lagrange function

$$\mathcal{L}(\vec{a}, \vec{S}, \Lambda) = \sum_{n=0}^{N-1} \phi(\vec{s}_{n+1}, a_n) + \sum_{n=0}^{N-1} \vec{\lambda}_n^T\, \Delta\vec{s}_n + \sum_{n=0}^{N-1} \vec{\lambda}_n^T\, (-A\,\vec{s}_n - b\,a_n),$$

where $\Delta\vec{s}_n = \vec{s}_{n+1} - \vec{s}_n$, $n = 0, 1, \ldots$ is the forward difference operator, while

$$\vec{\Lambda} \stackrel{def}{=} [\vec{\lambda}_0^T, \vec{\lambda}_1^T, \ldots, \vec{\lambda}_{N-1}^T]^T$$

is a supervector of the Lagrange multipliers $\vec{\lambda}_n$, which are $d_s \times 1$ vectors to be selected. Similarly, $\vec{S}$ is also supervector defined as follows:

$$\vec{S} \stackrel{def}{=} [\vec{s}_1^T, \vec{s}_2^T, \ldots, \vec{s}_N^T]^T.$$

From the summation by parts formula we obtain

$$\sum_{n=0}^{N-1} \vec{\lambda}_n^T \, \Delta \vec{s}_n = [\vec{\lambda}_{N-1}^T \, \vec{s}_N - \vec{\lambda}_0^T \, \vec{s}_0] - \sum_{n=1}^{N-1} (\vec{\lambda}_n^T - \vec{\lambda}_{n-1}^T) \, \vec{s}_n. \tag{7.26}$$

Observe that the second summand in the brackets vanishes, due to zero initial conditions for $\vec{s}_0$, which yield

$$\sum_{n=0}^{N-1} \vec{\lambda}_n^T \, \Delta \vec{s}_n = \vec{\lambda}_{N-1}^T \, \vec{s}_N - \sum_{n=1}^{N-1} \Delta \vec{\lambda}_{n-1}^T \, \vec{s}_n, \tag{7.27}$$

where $\Delta \vec{\lambda}_{n-1} = (\vec{\lambda}_n - \vec{\lambda}_{n-1})$. Collecting together the terms containing both $\vec{\lambda}_n$ and $\vec{s}_n$, we obtain

$$\mathcal{L}(\vec{a}, \vec{S}, \vec{\Lambda}) = \sum_{n=1}^{N} [\phi(\vec{s}_n, a_{n-1}) - \vec{\lambda}_{n-1}^T \, b \, a_{n-1}] + \vec{\lambda}_{N-1}^T \, \vec{s}_N - \\ - \sum_{n=1}^{N-1} [\Delta \vec{\lambda}_{n-1}^T + \vec{\lambda}_n^T \, A] \, \vec{s}_n. \tag{7.28}$$

Observe that in the above formula we can start the summation of the second sum from $n = 1$, since $\vec{s}_0 = \bar{0}$, while in the first one the shift of the variable was made.

From the Kuhn–Tucker theory (see e.g., [89]) it is known that if $\vec{a}^*$ and $\vec{S}^*$ minimizes $J(\vec{a})$ under constraints (7.24), then there exists nonzero vector $\vec{\Lambda}^*$ such that the following conditions hold for the gradients of $\mathcal{L}$

$$\nabla_{\vec{a}} \, \mathcal{L}(\vec{a}, \vec{S}, \vec{\Lambda})\Big|_{(\vec{a}^*, \vec{S}^*, \vec{\Lambda}^*)} = \bar{0}, \quad \nabla_{\vec{S}} \, \mathcal{L}(\vec{a}, \vec{S}, \vec{\Lambda})\Big|_{(\vec{a}^*, \vec{S}^*, \vec{\Lambda}^*)} = \bar{0}, \\ \nabla_{\vec{\Lambda}} \, \mathcal{L}(\vec{a}, \vec{S}, \vec{\Lambda})\Big|_{(\vec{a}^*, \vec{S}^*, \vec{\Lambda}^*)} = \bar{0}, \tag{7.29}$$

where $\bar{0}$ stands for vectors of zeros of the appropriate dimensions. For linear constraints (7.24) and strictly convex objective function these conditions are also sufficient for the optimality of $\vec{a}^*$, $\vec{S}^*$.

From the necessity of (7.29) it follows that $(\vec{a}^*, \vec{S}^*, \vec{\Lambda}^*)$ is a solution of the following set of equations that are obtained by calculating the gradients of $\mathcal{L}(\vec{a}, \vec{S}, \vec{\Lambda})$ and equating them to zero

$$\frac{\partial \, \phi(\vec{s}_{n+1}, a_n)}{\partial \, a_n} - \vec{\lambda}_n^T \, b = 0, \quad n = 0, 1, \ldots, (N-1), \tag{7.30}$$

$$\vec{\lambda}_n - \vec{\lambda}_{n-1} = -A^T \, \vec{\lambda}_n + \nabla_{\vec{s}} \phi(\vec{s}, a_{n-1})\Big|_{\vec{s}=\vec{s}_n}, \tag{7.31}$$

which is solved backward in time $n = (N-1), (N-2), \ldots, 1$ with the final condition

$$\vec{\lambda}_{N-1} = -\nabla_{\vec{s}}\phi(\vec{s}, a_{N-1})\Big|_{\vec{s}=\vec{s}_N}. \tag{7.32}$$

The third set of equations in (7.29) clearly yields the process equations:

$$\vec{s}_{n+1} - \vec{s}_n \;=\; A\,\vec{s}_n + b\,a_n, \quad \vec{s}_0 = \vec{0}, \tag{7.33}$$

that are solved forward in time $n = 0,\ 1, \ldots,\ (N-1),$.

At this stage of derivations it is customary to formulate the necessary optimality conditions as Pontriagin's maximum (minimum) principle (see [18, 115]). Here, we take another route, since the above set of equations is well suited for an iterative algorithm approximating its solution with

$$\nabla_{\vec{a}}\,\mathcal{L}(\vec{a}, \vec{S}, \vec{\Lambda}) = \left[\frac{\partial\,\phi(\vec{s}_1,\, a_0)}{\partial\, a_0} - \vec{\lambda}_0^T\, b,\ \ldots,\ \frac{\partial\,\phi(\vec{s}_N,\, a_{N-1})}{\partial\, a_{N-1}} - \vec{\lambda}_{N-1}^T\, b\right]^T \tag{7.34}$$

as an update of an approximate solution $\vec{a}(k)$ at $k$th iteration.

**Computational Algorithm for Optimizing Decision Sequences**

**Step 0** Select $\epsilon > 0$ for stopping condition and the starting sequence $\vec{a}(0)$. Set $k = 0$.

**Step 1** Compute the process response $\vec{s}_n(k)$, by iterating

$$\vec{s}_{n+1}(k) - \vec{s}_n(k) \;=\; A\,\vec{s}_n(k) + b\,a_n(k), \quad \vec{s}_0(k) = \vec{0}, \tag{7.35}$$

forward in time $n = 0,\ 1, \ldots,\ (N-1)$.

**Step 2** Compute the adjoint variables

$$\vec{\lambda}_n(k) - \vec{\lambda}_{n-1}(k) \;=\; -A^T\,\vec{\lambda}_n(k) + \nabla_{\vec{s}}\phi(\vec{s}, a_{n-1}(k))\Big|_{\vec{s}=\vec{s}_n(k)}, \tag{7.36}$$

by iterating backward in time $n = (N-1),\ (N-2), \ldots,\ 1$ with the final condition

$$\vec{\lambda}_{N-1}(k) = -\nabla_{\vec{s}}\phi(\vec{s}, a_{N-1}(k))\Big|_{\vec{s}=\vec{s}_N(k)}. \tag{7.37}$$

**Step 3** Compute the direction of search $\vec{d}(k)$ according to (7.34),

$$\vec{d}(k) \;=\; \nabla_{\vec{a}}\,\mathcal{L}(\vec{a}, \vec{S}(k), \vec{\Lambda}(k))\Big|_{\vec{a}=\vec{a}(k)},$$

then update the whole decision sequence

$$\vec{a}(k+1) = \vec{a}(k) - \alpha_k\,\vec{d}(k). \tag{7.38}$$

If $||\vec{d}(k)|| < \epsilon$, then STOP and provide $\vec{a}(k+1)$ as the result, otherwise set $k := k+1$ and go to Step 1.

Step size $\alpha_k > 0$ in (7.38) can be either a small constant or it can be selected as the minimizer in search direction $\vec{d}(k)$. Sufficient conditions for covergence of this algorithm are typical for gradient type methods (see e.g., [89]). One can also try to modify $\vec{d}(k)$'s in the spirit of the quasi-Newton methods, but for large $N$ only scaling the elements of $\vec{d}(k)$.'s seems to be applicable.

The above computational algorithm serves as an inspiration for the following learning approach.

**Learning Algorithm for Optimizing Decision Sequences**

**Step 0** Set the counter of passes $k = 0$. Select the starting sequence $\vec{\tilde{a}}(0)$ as well as the step size $\alpha_k > 0$ for updates, which is either a small constant or selected in such a way that

$$\lim_{k\to\infty} \alpha_k = 0, \quad \sum_{k=1}^{\infty} \alpha_k = \infty, \quad \sum_{k=1}^{\infty} \alpha_k^2 < \infty. \tag{7.39}$$

**Step 1** Apply decision sequence $\vec{\tilde{a}}(k)$ to the process, observe and store its states[3]: $\vec{\tilde{s}}_n(k)$, $n = 1, 2, \ldots, N$.

**Step 2** Compute the adjoint variables using observations $\vec{\tilde{s}}_n(k)$, $n = 1, 2, \ldots, N$ instead of state variables from the model, i.e.,

$$\vec{\tilde{\lambda}}_n(k) - \vec{\tilde{\lambda}}_{n-1}(k) = -A^T \vec{\tilde{\lambda}}_n(k) + \nabla_{\vec{s}}\phi(\vec{s}, a_{n-1}(k))\Big|_{\vec{s}=\vec{\tilde{s}}_n(k)}, \tag{7.40}$$

by iterating backward in time $n = (N-1), (N-2), \ldots, 1$ with the final condition

$$\vec{\tilde{\lambda}}_{N-1}(k) = -\nabla_{\vec{s}}\phi(\vec{s}, \tilde{a}_{N-1}(k))\Big|_{\vec{s}=\vec{s}_N(k)}. \tag{7.41}$$

**Step 3** Compute the direction of search $\vec{\tilde{d}}(k)$ according to (7.34),

$$\vec{\tilde{d}}(k) = \nabla_{\vec{a}}\, \mathcal{L}(\vec{a}, \vec{\tilde{S}}(k), \vec{\tilde{\Lambda}}(k))\Big|_{\vec{a}=\vec{\tilde{a}}(k)},$$

then update the whole decision sequence

$$\vec{\tilde{a}}(k+1) = \vec{\tilde{a}}(k) - \alpha_k\, \vec{\tilde{d}}(k). \tag{7.42}$$

Set $k := k + 1$ and go to Step 1.

In Step 3, $\vec{\tilde{S}}(k)$ and $\vec{\tilde{\Lambda}}(k)$ are composed of $\vec{\tilde{s}}_n(k)$'s and $\vec{\tilde{\lambda}}_n(k)$'s, respectively.

The main difference between this learning version and the previous computational algorithm is in that the former the model is partly replaced by acting and observing

---

[3] If process states are not available for observations, one can apply a state observer to reconstruct the states from observations of the process output.

a real process. On the other hand, the model is still needed for computing the adjoint variables, which—in turn—are necessary for evaluating the search directions $\vec{d}(k)$'s. One possible advantage of such a model-supported approach is that it can be more robust to model inaccuracy than a fully model-based approach. The rationale standing behind this statement is the following: if the model is not exact, but observations of the process states are only slightly corrupted by random errors, then one can expect that $\vec{d}(k)$'s are still descent directions, although their computations are based on an inexact model. More detailed analysis of the robustness of this approach is outside the scope of this book.

Notice that—as opposed to the computational algorithm—the learning algorithm does not have a stopping condition. This is done intentionally to keep its learning abilities for possible inaccuracies, e.g., in the initial conditions for each pass. If a decreasing sequence of the step lengths $\alpha_k$' is used, then it can be necessary to to restart the learning process from time to time to keep the learning abilities of the algorithm. The discussion on using adaptive choice of the step length and scaling provided at the end of Sect. 6.4.1 is also relevant here.

In the next section an example of the computational version of this algorithm is provided for the quadratic objective function.

## 7.4 Derivation of the Learning Algorithm

Let us consider the discrete time model, during the $k$th pass of the learning algorithm. Generally $k = 1, 2, \ldots$ are used as numbers of subsequent passes.

$$\vec{s}_{n+1}(k) = A\vec{s}_n(k) + \vec{b}a_n(k),\ n = 0, \ldots, N\ \vec{s}_0(k) = \vec{s}_0 \tag{7.43}$$

where (defined in 2.1.1) $\vec{s}_{n+1}(k)$ is the discrete state in $n$th state during $k$th pass. Obviously the initial condition is given by $\vec{s}_0$ and does not change in each pass of the learning algorithm.[4] $A$ is matrix $m \times m$, $\vec{b}$ is an $m$-element vector forming a discrete-time model. Particular decisions are denoted by $m$-dimensional vector $a_n(k)$ where, as previously, $k$ is the pass number and $n$ is the discrete time of action taken.

We define quality functional $J(\vec{a})$

$$J(\vec{a}) = \sum_{i=0}^{N} \left[ \frac{1}{2} (\vec{s}_i^* - \vec{s}_i)^T (\vec{s}_i^* - \vec{s}_i) + r a_i^2 \right] \tag{7.44}$$

as a sum of squared differences between the expected state and achieved state and weighted action cost understood in energy terms. The weighting factor is the term $r$.

[4] In a general case it can change from pass to pass.

We formulate the optimization problem in the form

$$\min_{\vec{a}} J(\vec{a}) \tag{7.45}$$

with constraint

$$\vec{s}_{n+1} = A\vec{s}_n + \vec{b}^T a_n, n = 0, \dots, \vec{s}(0) = s_0$$

This problem is similar to the discrete linear-quadratic control problem, which has a well-known solution using the Riccati equation.

Our aim is to improve $a_n(k)$ using an iterative algorithm.

$$a_{.}(k+1) = \Psi_k(\vec{s}(k), a(k)) \tag{7.46}$$

From such an algorithm we require that:

$$\lim_{k\to\infty} J(a(k)) = J(a^*)$$

$$\lim_{k\to\infty} ||a(k) - a^*|| = 0$$

$$\lim_{k\to\infty} ||a(k) - a^*||_d = 0$$

The hamiltonian for this problem has the form

$$H(\vec{s}_k, a_k, \psi_k) = (\vec{s}_k^* - \vec{s}_k)^T(\vec{s}_k^* - \vec{s}_k) + ra_k^2 + \psi^T[A\vec{s}_k + \vec{b}^T a_k] \tag{7.47}$$

and

$$\psi_i = (I - A)\psi_{k+1} + (\vec{s}_k^* - \vec{s}_k),\ \psi_N = 0 \tag{7.48}$$

The locally steepest descent direction can be easily calculated using the gradient of Hamiltonian

$$F(\vec{s}, a, \psi) = \nabla_a H = 2ra + \vec{b}^T a \tag{7.49}$$

If $\vec{s}^*$ and $a^*$ are the solution to the problem stated previously then the following holds

$$F^* = F(\vec{s}^*, a^*, \psi^*) = 0 \tag{7.50}$$

$$\vec{s}_{n+1}^*(k) = As_n^*(k) + Ba_n^*(k),\ n = 0, \dots, N, \quad \vec{s}_0(k) = \vec{s}_0 \tag{7.51}$$

$$\psi_i^* = (I - A)\psi_{i+1}^*,\ \psi_N = 0 \tag{7.52}$$

The following update is proposed

$$a_n(k+1) = a_n(k) + \gamma F_n(k) \tag{7.53}$$

where

$$F_n(k) = (2ra_n(k) + \vec{b}^T \psi_n) \tag{7.54}$$

$$a_n(k+1) = (1 - 2\gamma r)a_n(k) + \gamma \vec{b}^T \psi_n \tag{7.55}$$

The $\psi_n$ is calculated by back-stepping (7.52)

The learning parameter $\gamma$ has to be chosen in such a way that $J(a_n)$ converges. Let us expand the $J$ into the Taylor series in the following way.

$$J(a_n - \gamma \mathcal{A}) = J(a_n) + \gamma \nabla J^T \mathcal{A} + 2r\frac{\gamma^2}{2}\mathcal{A}^T \mathcal{A} \tag{7.56}$$

The obvious selection of action $\mathcal{A}$ is the steepest descent $\mathcal{A} = -F_n$, by substitution we obtain

$$J(a_{n+1}) = J(a_n - \gamma F_n) = J(a_n) - \gamma F_n^T F_n + (2r)\frac{\gamma}{2} F_n^T F_n \tag{7.57}$$

Convergence is assured if $J(a(k+1)) < J(a(k))$, when $\gamma < \frac{1}{r}$ or for $\nu > 0$ $\gamma = \frac{1}{r+\nu}$, since $J(a(k+1))$ is monotonic and bounded from below by 0, thus the sequence $J(a(k)$ is convergent. The fact that it is convergent to $J(a^*)$ can be proven in the same way as in the paper [140].

## 7.5 Pass to Pass Learning

The learning process can also be applied on process interaction from one pass to another. In this case we use the direct observations of a system instead of results from the model.

This would compensate for inaccuracies and simplification of a model and possible fluctuations and drift.

**Step 0** Select $a(0)$ possibly by using the process model, set $k = 0$, select desired accuracy $\varepsilon > 0$.
**Step 1** Apply $a(k)$ to a real system, store $\hat{s(k)}$.
**Step 2** Calculate $\psi(k)$ by backtracking.
**Step 3** Calculate $\hat{F}(k) = 2ra(k) + \vec{b}^T a(k)$. If $\max_i |\hat{F}(k)_i| > \varepsilon$ set $k = k + 1$ and go to Step 1.
**Step 4** Update $a(k+1) = a(k) - \gamma\hat{F}(k)$, $k = k + 1$ and go to step 1.

# Chapter 8
# Learning from Image Sequences

## 8.1 Motivation and Aims

Many processes are divided into a certain number of stages. The proper diagnostic decisions can be taken on intermediate process stages and only their junction provides a proper result.

Machine vision is one of the main tools in diagnostics. It allows for quick assessment—a short time required for image acquisition—of the process. The images can be taken at different but subsequent stages of the process. It requires classifying image sequences from all the diagnostic stages as one entity.

A method of classifying the whole image sequence as proper or improper (conforming or non-conforming) without imposing assumptions on probability distributions of images (nonparametric approach) is proposed in this chapter.

An example at the end of this chapter comes from an industrial process, but the proposed approach has much wider potential applications, e.g., for assessing whether a surgeon properly accomplish all the stage of a laparoscopic surgery.

## 8.2 Proposed Classifier

Let us assume that $X = \{X_1, X_2, \ldots, X_m\}$ be a sequence of random matrices representing gray-level images. These images should be acquired in similar conditions e.g. with the same lighting, same hardware and without much movement. General rules used in image processing apply here.

A learning sequence (of sequences) is given with labels: $L^l$—"BAD" , "OK" corresponding to previously-mentioned images forming pairs $(X_1, L^1), \ldots (X_m, L^m)$.

We intend to construct the classifier $C_N(X) \rightarrow \{$"BAD" , "OK" $\}$.



W. Rafajłowicz, *Learning Decision Sequences For Repetitive Processes—Selected Algorithms*, Studies in Systems, Decision and Control 401,
https://doi.org/10.1007/978-3-030-88396-6_8

The two classes are defined as follows:

**"OK"** —conforming,
**"BAD"** —nonconforming result.

A priori class probabilities and distributions—exist, but are considered unknown (nonparametric setting).

Many approaches to this problem are possible.

One can extract features vectors $F_i$ from images $X_i, i = 1, 2, \ldots m$, form sequence $F = \{f_1, f_2, \ldots, f_m\}$ and use it for learning and recognition. This is a very laborious process, but sometimes the only possible approach.

In a new, nonparamtetric approach we use images $X_i$ , i=1,2,...m directly for learning and recognition. In this method, only fast and easy pre-processing is allowed. The method is applicable for simple, well-defined processes.

In [130] a parametric approach was proposed, assuming that we have a matrix normal distribution (MND) with p.d.f.

$$f(X) = \frac{1}{c}\exp\left[-\frac{1}{2}\mathrm{tr}(U^{-1}(X-M)V^{-1}(X-M)^T)\right] \tag{8.1}$$

where $c$ is the normalization constant, $M$ is the mean matrix, $U$ and $V$ are covariance matrices for rows and columns of the image X.

The matrices are as follows: $X$ is an $m \times n$ image, $U$ is $m \times m$, $V$ is $n \times n$ (not $m \cdot n \times m \cdot n$ as in general multivariate Gaussian distributions).

Let $M = \{M_1, M_2, \ldots, M_m\}$ be a sequence of matrices of the same dimensions as $X_i$'s.

As a distance $\Omega(X, M)$ between sequences of images $X$ and $M$ one can take: $\Omega(X, M) = \max_{1 \le i \le m} \rho_i(M_i, X_i)$ , where $\rho_i(M_i, X_i)$ is the distance between single images $M_i$ , $X_i$ . Alternatively,

$$\Omega(X, M) = \sum_{1 \le i \le m} \rho_i(M_i, X_i) \tag{8.2}$$

is further considered, with $\rho_i$ defined as the generalized Frobenius norm of $(M_i - X_i)$:

$$\rho_i(M_i, X_i) = \mathrm{tr}\left[A_i \cdot (M_i - X_i) \cdot B_i \cdot (M_i - X_i)^T\right] \tag{8.3}$$

where $A_i$ and $B_i$—are $A_i = \alpha_i \cdot I_m$ , $B_i = \beta_i \cdot I_n$ . The $\alpha_i, \beta_i$ are scalars or symmetric and positive definite matrices, selected by the user (one can take $A_i$ and $B_i$ as the inverses of the rows and columns covariance matrix as in the MND case, but estimation requires long learning sequences—thousands of images even for m = n = 10).

Further we take:

$$\alpha_i = ||vec(X_i) - E[vec(X_i)]||^{-1}, \tag{8.4}$$

$$\beta_i = ||vec(M_i)] - E[vec(M_i)]||^{-1}, \tag{8.5}$$

where $vec(X_i)$—$m\,n$ vectors from stacked columns of $X_i$ , $E$—the expectation—in practice, replaced by the empirical mean and $||.||$ is the Euclidean norm in $R^{m \cdot n}$ .

Let $M^O = M_1^O, M_2^O, \ldots, M_m^O$ be a sequence of template images for the "OK" class, obtained from the learning sequence e.g. as pixel-by-pixel means (or medians or …) of images from the "OK" class in the learning sequence.

Let $M^B = M_B^1, M_B^2, \ldots, M_B^m$ be a sequence of template images for "OK" class, obtained from the learning sequence as above.

Proposed classifier: classify new sequence X as "OK" if

$$\widehat{p_o} \cdot \Omega(X, M^O) < \widehat{p_B} \cdot \Omega(X, M^B),$$

as "BAD" otherwise, where $\widehat{p_O}$ and $\widehat{p_B}$ are empirical frequences of sequences labeled as "OK" and "BAD" in the learning sequence of sequences. This mimicks the Bayes classifier, but we do not make the assumptions that are typical for its derivation.

Faster version—maximum correlation sum classifier for sequences of images. For scalar weights $\alpha_i$, $\beta_i$ the classifier can be simplified by replacing

$$\begin{aligned} \rho_i(M_i, X_i) &= \alpha_i \cdot \beta_i \cdot \mathrm{tr}[(M_i - X_i) \cdot (M_i - X_i)^T \\ &= \alpha_i \cdot \beta_i \cdot [(M_i - X_i)^T \cdot vec[(M_i - X_i)^T] \end{aligned} \tag{8.6}$$

by the correlations:

$$\begin{aligned} \widehat{\rho_o}(M_i, X_i) = -2 \cdot \alpha_i \cdot \beta_i \cdot [(vec(M)_i^T - E[vec(M)_i^T] \\ \cdot vec(X)_i^T - E[vec(X)_i^T], \end{aligned} \tag{8.7}$$

since the terms that are quadratic in $M_i$ and $X_i$ are almost constant (see [51]).

**Faster classifier**: classify new sequence $X$ as "OK" if $\widehat{\rho_o}\Omega(X, M^O) < \widehat{\rho_B}\Omega(X, M^B)$

where $\Omega(X, M^O)$ is obtained from $\Omega(X, M^B)$ by replacing $\rho_i(M_i, X_i)$ by $\widehat{\rho_i}(M_i, X_i)$ There is no fear of summing up correlations, since negative ones are witness against a given class.

When images in $X, M^O, M^B$ are binary, then the inner products in reduce to boolean AND operations that can be performed parallelly on GPU and their results just counted. Additionally, $M^O$ , $M^B$ can be pre-computed and stored.

## 8.3 Example

Laser cladding is an additive 3D printing process that uses metallic powder. During the printing a laser head moves back and forth, melting the powder and thus creating a shape (see [138, 139]).

A constant laser power is not enough to obtain a prescribed shape of a printed 3D body. The reason is in that the laser head has to turn back at the ends of the body. Hence, the laser head has to slow down near the end, then to stop and speed up again.

This results in adding too much metallic powder near the end points, which—in turn—leads to forming undesirable bulb-like shapes.

The learning sequence consists of 449 images (56—"Bad", 393—"OK") (Fig. 8.1).

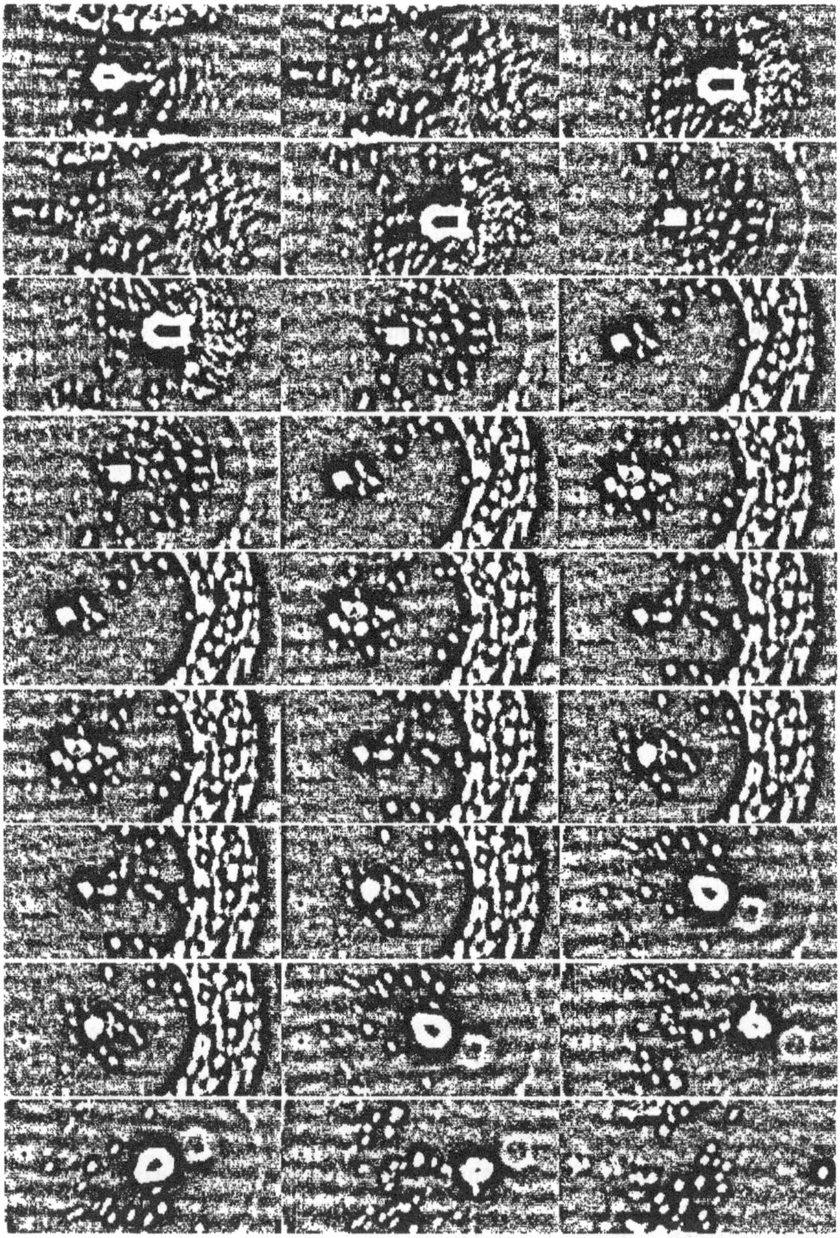

**Fig. 8.1** Example of binary images from laser cladding process

A rather large class imbalance is present. This is typical for most industrial processes that are usually good. The nearest mean method is relatively robust (see [129]).

The testing sequence consists of 449 images, including: 54 from class "BAD", 395 from class "OK"

The results are as follows

| | | |
|---|---|---|
| False positive (FP) | 11 | ("BAD" as "OK") |
| False negative (FN) | 0 | ("OK" as "BAD") |
| True positive (TP) | 395 | ("OK" as "OK") |
| True negative (TN) | 43 | ("BAD" as "BAD") |

The resulting accuracy, defined as

$$100\%(TP + TN)/(FP + FN + TP + TN) = 97.6\%, \tag{8.8}$$

is very good, but when classes are imbalanced the accuracy may be artificially high (classifying improperly examples from the minority class).

The sensitivity, defined as 100% TP/(TP+FN) = 100% but TN and FP cases are neglected.

The precision, defined as 100% TP/(TP+FP)=97.3% but TN and FN cases are neglected.

The F1 score = 2.0 $\cdot$ $Prec \cdot Sens/(Prec + Sens) = 98.6\%$ but TN neglected.

Matthews Correlation Coefficient (MCC)

$$MCC = \frac{TP\,TN - FP\,FN}{\sqrt{(TP + FP) \cdot (TP + FN) \cdot (TN + FP) \cdot (TN + FN)}} = 0.88 \tag{8.9}$$

Includes all cases (TP, TN, FP, FN)

The interpratation of MCSS is as follows: a classifier with MCC close $+/-$ 1 has a good prediction, one with MCC close to 0 is comparable to tossing a coin.

The proposed classifier of image sequences provided high performance according to all the widely-used measures of quality.

In the real world, application of laser control requires specific decisions rules. For example, the rules can be as follows:

**3 images classified as "OK"** keep the laser power high
**3 images classified as "BAD"** reduce the laser power

This has to be applied repetitively—twice for each pass (near the left and the right end). More subtle classification of "BAD" cases allows for a more finely-tuned reduction in laser power.

# Bibliography


1. Ahn, H.-S., Chen, Y., Moore, K.L.: Iterative learning control: brief survey and categorization. IEEE Trans. Syst. Man Cybern. Part C Appl. Rev. **37**(6), 1099 (2007)
2. Akaike, H.: A new look at the statistical model identification. IEEE Trans. Autom. Control **19**(6), 716–723 (1974)
3. Al-Dabbagh, R.D., Neri, F., Idris, N., Baba, M.S.: Algorithmic design issues in adaptive differential evolution schemes: review and taxonomy. Swarm Evol. Comput. **43**, 284–311 (2018)
4. Annas, S., Pratama, M.I., Rifandi, M., Sanusi, W., Side, S.: Stability analysis and numerical simulation of SEIR model for pandemic COVID-19 spread in Indonesia. Chaos, Solitons Fractals **139**, 110072 (2020)
5. Arabas, J., Szczepankiewicz, A., Wroniak, T.: Experimental comparison of methods to handle boundary constraints in differential evolution. In: International Conference on Parallel Problem Solving from Nature, pp. 411–420. Springer (2010)
6. Aschemann, H., Rauh, A.: An integro-differential approach to control-oriented modelling and multivariable norm-optimal iterative learning control for a heated rod. In: 2015 20th International Conference on Methods and Models in Automation and Robotics (MMAR), pp. 447–452 (2015)
7. Bach, F., Moulines, E.: Non-strongly-convex smooth stochastic approximation with convergence rate o(1/n). In: Burges, C.J.C., Bottou, L., Welling, M., Ghahramani, Z., Weinberger, K.Q. (eds.) Advances in Neural Information Processing Systems, vol. 26, pp. 773–781. Curran Associates, Inc. (2013)
8. Barron, A., Rissanen, J., Bin, Y.: The minimum description length principle in coding and modeling. IEEE Trans. Inf. Theory **44**(6), 2743–2760 (1998)
9. Bello, I., Pham, H., Le, Q.V., Norouzi, M., Bengio, S.: Neural combinatorial optimization with reinforcement learning. In: 5th International Conference on Learning Representations, ICLR 2017 - Workshop Track Proceedings, pp. 1–15 (2019)
10. Bertsekas, D.P.: Approximate dynamic programming (2008)
11. Biedrzycki, R., Arabas, J., Jagodziński, D.: Bound constraints handling in differential evolution: an experimental study. Swarm Evol. Comput. **50**, 100453 (2019)
12. Blum, J.R.: Approximation methods which converge with probability one. Ann. Math. Stat., pp. 382–386 (1954)




W. Rafajłowicz, *Learning Decision Sequences For Repetitive Processes—Selected Algorithms*, Studies in Systems, Decision and Control 401,
https://doi.org/10.1007/978-3-030-88396-6

13. Blum, J.R.: Multidimensional stochastic approximation methods. Ann. Math. Stat., pp. 737–744 (1954)
14. Bocewicz, G., Banaszak, Z.A.: Declarative approach to cyclic steady state space refinement: periodic process scheduling. Int. J. Adv. Manuf. Technol. **67**(1–4), 137–155 (2013)
15. Bock, W., Adamik, B., Bawiec, M., Bezborodov, V., Bodych, M., Burgard, J.P., Goetz, T., Krueger, T., Migalska, A., Pabjan, B. et al.: Mitigation and herd immunity strategy for covid-19 is likely to fail. medRxiv (2020)
16. Bolder, J., Kleinendorst, S., Oomen, T.: Data-driven multivariable ILC: enhanced performance by eliminating L and Q filters. Int. J. Robust Nonlinear Control **28**(12), 3728–3751 (2018)
17. Boltyanski, V.G., Poznyak, A.: The Robust Maximum Principle: Theory and Applications. Springer Science & Business Media (2011)
18. Boltyanski, V.G.: Optimal Control of Discrete Systems. Halsted Press, Sydney (1978)
19. Borkar, V.S.: Asynchronous stochastic approximations. SIAM J. Control. Optim. **36**(3), 840–851 (1998)
20. Bortz, D.M., Kelley, C.T.: The Simplex Gradient and Noisy Optimization Problems, pp. 77–90. Birkhäuser, Boston (1998)
21. Bošković, B., Greiner, S., Brest, J., Žumer, V.: A differential evolution for the tuning of a chess evaluation function. In: 2006 IEEE Congress on Evolutionary Computation, CEC 2006, pp. 1851–1856 (2006)
22. Bottou, L., Curtis, F.E., Nocedal, J.: Optimization methods for large-scale machine learning. Siam Rev. **60**(2), 223–311 (2018)
23. Bożejko, W., Gnatowski, A., Niżyński, T., Affenzeller, M., Beham, A.: Local optima networks in solving algorithm selection problem for tsp. In: International Conference on Dependability and Complex Systems, pp. 83–93. Springer (2018)
24. Bristow, D., Tharayil, M., Alleyne, A.G. et al.: A survey of iterative learning control. IEEE Control Syst. **26**(3), 96–114 (2006)
25. Byrd, R.H., Hansen, S.L., Nocedal, J., Singer, Y.: A stochastic quasi-newton method for large-scale optimization. SIAM J. Optim. **26**(2), 1008–1031 (2016)
26. Celik, E., Gul, M., Aydin, N., Gumus, A.T., Guneri, A.F.: A comprehensive review of multi criteria decision making approaches based on interval type-2 fuzzy sets. Knowl.-Based Syst. **85**, 329–341 (2015)
27. Chakraborty, U.K.: Advances in Differential Evolution, vol. 143. Springer, Berlin (2008)
28. Chau, M., Fu, M.C.: An Overview of Stochastic Approximation, pp. 149–178. Springer, New York (2015)
29. Chau, M., Fu, M.C.: An overview of stochastic approximation. Handbook of Simulation Optimization, pp. 149–178, Springer, Berlin (2015)
30. Chen, H.F., Duncan, T.E., Pasik-Duncan, B.: A Kiefer-Wolfowitz algorithm with randomized differences. IEEE Trans. Autom. Control **44**(3), 442–453 (1999)
31. Chu, B., Owens, D.H., Freeman, C.T.: Iterative learning control with predictive trial information: convergence, robustness, and experimental verification. IEEE Trans. Control. Syst. Technol. **24**(3), 1101–1108 (2015)
32. Cichy, B., Gałkowski, K., Rogers, E.: Iterative learning control for spatio-temporal dynamics using Crank-Nicholson discretization. Multidimension. Syst. Signal Process. **23**(1–2), 185–208 (2012)
33. Cichy, B., Gałkowski, K., Rogers, E., Kummert, A.: An approach to iterative learning control for spatio-temporal dynamics using nd discrete linear systems models. Multidimension. Syst. Signal Process. **22**(1–3), 83–96 (2011)
34. Costello, S., François, G., Srinivasan, B., Bonvin, D.: Modifier adaptation for run-to-run optimization of transient processes. IFAC Proc. Vol. **44**(1), 11471–11476 (2011)
35. Deudon, M., Cournut, P., Lacoste, A., Adulyasak, Y., Rousseau, L.-M.: Learning heuristics for the tsp by policy gradient. In: van Hoeve, W.-J. (ed.) Integration of Constraint Programming, Artificial Intelligence, and Operations Research, pp. 170–181. Springer International Publishing, Cham (2018)

36. Devroye, L., Györfi, L., Lugosi, G.: A Probabilistic Theory of Pattern Recognition, vol. 31. Springer Science & Business Media (2013)
37. Dippon, J., Fabian, V.: Stochastic approximation of global minimum points. J. Stat. Plan. Inference **41**(3), 327–347 (1994)
38. Duarte, F.F., Lau, N., Pereira, A., Reis, L.P.: A survey of planning and learning in games. Appl. Sci. (Switzerland) **10**(13) (2020)
39. Duchi, J., Hazan, E., Singer, Y.: Adaptive subgradient methods for online learning and stochastic optimization. J. Mach. Learn. Res. **12**(7) (2011)
40. Ehrgott, M.: Multicriteria Optimization, vol. 491. Springer Science & Business Media (2005)
41. Fabian, V.: Stochastic approximation of minima with improved asymptotic speed. Ann. Math. Stat., pp. 191–200 (1967)
42. Feoktistov, V.: Differential Evolution. Springer, Berlin (2006)
43. Fischer, H.: Automatic differentiation: parallel computation of function, gradient, and hessian matrix. Parallel Comput. **13**(1), 101–110 (1990)
44. Fletcher, R., Leyffer, S.: Nonlinear programming without a penalty function. Math. Program. **91**(2), 239–269 (2002)
45. Fogel, D.B., Hays, T.J., Hahn, S.L., Quon, J.: A self-learning evolutionary chess program. Proc. IEEE **92**(12), 1947–1954 (2004)
46. Fu, M.C. (ed.): Springer, New York (2015)
47. Galar, R.: Handicapped individua in evolutionary processes. Biol. Cybern. **53**(1), 1–9 (1985)
48. Galar, R.: Evolutionary search with soft selection. Biol. Cybern. **60**(5), 357–364 (1989)
49. Ghosh, A., Tsutsui, S.: Advances in Evolutionary Computing: Theory and Applications. Springer Science & Business Media (2012)
50. Ghosh, S., Das, S., Vasilakos, A.V., Suresh, K.: On convergence of differential evolution over a class of continuous functions with unique global optimum. IEEE Trans. Syst., Man, Cybern., Part B: Cybern. **42**(1), 107–124 (2012)
51. Gonzales, R., Woods, R.: Digital Image Processing, 3rd edn. (2008)
52. Greblicki, W.: Learning to recognize patterns with a probabilistic teacher. Pattern Recogn. **12**(3), 159–164 (1980)
53. Greblicki, W., Pawlak, M.: Nonparametric System Identification, vol. 1. Cambridge University Press, Cambridge (2008)
54. Groba, C., Sartal, A., Vázquez, X.H.: Solving the dynamic traveling salesman problem using a genetic algorithm with trajectory prediction: an application to fish aggregating devices. Comput. Oper. Res. **56**, 22–32 (2015)
55. Györfi, L., Kohler, M., Krzyżak, A., Walk, H.: A distribution-free theory of nonparametric regression, vol. 1. Springer, Berlin (2002)
56. Hladowski, L., Galkowski, K., Cai, Z., Rogers, E., Freeman, C.T., Lewin, P.L.: A 2d systems approach to iterative learning control with experimental validation. In: Proceedings of the 17th IFAC World Congress, Soeul, Korea, pp. 2832–2837 (2008)
57. Hladowski, L., Galkowski, K., Cai, Z., Rogers, E., Freeman, C.T., Lewin, P.L.: Experimentally supported 2d systems based iterative learning control law design for error convergence and performance. Control. Eng. Pract. **18**(4), 339–348 (2010)
58. Hou, Z., Jin, S.: Model Free Adaptive Control: Theory and Applications. CRC Press, Boca Raton (2013)
59. Zhongbo, H., Xiong, S., Qinghua, S., Fang, Z.: Finite Markov chain analysis of classical differential evolution algorithm. J. Comput. Appl. Math. **268**, 121–134 (2014)
60. Huo, B., Freeman, C.T., Liu, Y.: Model-free gradient iterative learning control for non-linear systems. IFAC-PapersOnLine **52**(29), 304–309 (2019)
61. Ilavarasi, K., Joseph, K.S.: Variants of travelling salesman problem: a survey. In: International Conference on Information Communication and Embedded Systems (ICICES2014), pp. 1–7. IEEE (2014)
62. Ingolfsson, A., Sachs, E.: Stability and sensitivity of an ewma controller. J. Qual. Technol. **25**(4), 271–287 (1993)

63. Jagodziński, D., Arabas, J.: A differential evolution strategy. In: 2017 IEEE Congress on Evolutionary Computation (CEC), pp. 1872–1876. IEEE (2017)
64. Jeyakumar, G., Shanmugavelayutham, C.: Convergence analysis of differential evolution variants on unconstrained global optimization functions. Int. J. Artif. Intell. Appl. **2**(2), 116–127 (2011)
65. Kacprzyk, J.: Multistage Fuzzy Control: A Model-based Approach to Fuzzy Control and Decision Making. Wiley, Hoboken (1997)
66. Kallel, L., Naudts, B., Rogers, A.: Theoretical Aspects of Evolutionary Computing. Springer Science & Business Media (2013)
67. Karlin, S.: A First Course in Stochastic Processes. Academic, Cambridge (2014)
68. Kiefer, J., Wolfowitz, J.: Stochastic estimation of the maximum of a regression function. Ann. Math. Stat. **23**, 462–466 (1952)
69. Kiefer, J., Wolfowitz, J., et al.: Stochastic estimation of the maximum of a regression function. Ann. Math. Stat. **23**(3), 462–466 (1952)
70. Kiwiel, K.C.: Methods of Descent for Nondifferentiable Optimization, vol. 1133. Springer, Berlin (2006)
71. Kluska, J.: Analytical Methods in Fuzzy Modeling and Control, vol. 241. Springer, Berlin (2009)
72. Knobloch, R., Mlýnek, J., Srb, R.: The classic differential evolution algorithm and its convergence properties. Appl. Math. **62**(2), 197–208 (2017)
73. Korbicz, J., Koscielny, J.M.: Modeling, Diagnostics and Process Control: Implementation in the DiaSter System. Springer Science & Business Media (2010)
74. Koronacki, J.: Random-seeking methods for the stochastic unconstrained optimization. Int. J. Control **21**(3), 517–527 (1975)
75. Koronacki, J.: Random-seeking methods for the stochastic unconstrained optimization. Int. J. Control **21**(3), 517–527 (1975)
76. Koronacki, J.: Some remarks on stochastic approximation methods, numerical techniques for stochastic systems, Edited by F. Archetti and M. Cugiani (1980)
77. Koronacki, J.: A stochastic approximation counterpart of the feasible direction method. Stat. Probab. Lett. **5**(6), 415–419 (1987)
78. Koval, V., Schwabe, R.: A law of the iterated logarithm for stochastic approximation procedures in d-dimensional euclidean space. Stoch. Process. Their Appl. **105**(2), 299–313 (2003)
79. Krawczyk, B., Triguero, I., García, S., Woźniak, M., Herrera, F.: Instance reduction for one-class classification. Knowl. Inf. Syst. **59**(3), 601–628 (2019)
80. Kushner, H.J., Yang, J.: Stochastic approximation with averaging and feedback: rapidly convergent "on-line" algorithms. IEEE Trans. Autom. Control **40**(1), 24–34 (1995)
81. Kushner, H.: Stochastic approximation: a survey. Wiley Interdiscip. Rev.: Comput. Stat. **2**(1), 87–96 (2010)
82. Kushner, H., Yin, G.G.: Stochastic Approximation and Recursive Algorithms and Applications, vol. 35. Springer Science & Business Media (2003)
83. Kushner, H.J., Clark, D.S.: Stochastic Approximation Methods for Constrained and Unconstrained Systems, vol. 26. Springer Science & Business Media (2012)
84. Lagarias, J.C., Reeds, J.A., Wright, M.H., Wright, P.E.: Convergence properties of the Nelder–Mead simplex method in low dimensions. SIAM J. Optim. **9**(1), 112–147 (1998)
85. Lake, B.M., Salakhutdinov, R., Tenenbaum, J.B.: Human-level concept learning through probabilistic program induction. Science **350**(6266), 1332–1338 (2015)
86. Leszek, R.: Computational Intelligence. Springer, Berlin (2008)
87. Liu, K., Chen, Y.Q., Zhang, T., Tian, S., Zhang, X.: A survey of run-to-run control for batch processes. ISA Trans. **83**, 107–125 (2018)
88. Lorentz, G.G.: Bernstein Polynomials. American Mathematical Society (2013)
89. Luenberger, D.G.: Optimization by Vector Space Methods. Wiley, Hoboken (1997)
90. Mandziuk, J.: Knowledge-Free and Learning-based Methods in Intelligent Game Playing, vol. 276. Springer, Berlin (2010)
91. Mavrovouniotis, M., Yang, S.: Ant colony optimization with immigrants schemes for the dynamic travelling salesman problem with traffic factors. Appl. Soft Comput. J. **13**(10), 4023–4037 (2013)

92. Michalewicz, Z., Schoenauer, M.: Evolutionary algorithms for constrained parameter optimization problems. Evol. Comput. **4**(1), 1–32 (1996)
93. Mohamed, A.W., Sabry, H.Z.: Constrained optimization based on modified differential evolution algorithm. Inf. Sci. **194**, 171–208 (2012)
94. Montgomery, D.C.: Design and Analysis of Experiments. Wiley, Hoboken (2017)
95. Montgomery, D.C.: Introduction to Statistical Quality Control. Wiley, Hoboken (2020)
96. Moore, K.L., Xu, J.-X.: Editorial: special issue on iterative learning control. Int. J. Control **73**(10) (2000)
97. Moulines, E., Bach, F.R.: Non-asymptotic analysis of stochastic approximation algorithms for machine learning. In: Shawe-Taylor, J., Zemel, R.S., Bartlett, P.L., Pereira, F., Weinberger, K.Q. (eds.) Advances in Neural Information Processing Systems, vol. 24, pp. 451–459. Curran Associates, Inc. (2011)
98. Moyne, J., Castillo, E.D., Hurwitz, A.M.: Run-to-run Control in Semiconductor Manufacturing. CRC Press, Boca Raton (2018)
99. Myers, R.H., Montgomery, D.C., Anderson-Cook, C.M.: Response Surface Methodology: Process and Product Optimization Using Designed Experiments. Wiley, Hoboken (2016)
100. Myers, R.H., Montgomery, D.C., Vining, G.G., Borror, C.M., Kowalski, S.M.: Response surface methodology: a retrospective and literature survey. J. Qual. Technol. **36**(1), 53–77 (2004)
101. Nazin, A.V., Polyak, B.T., Tsybakov, A.B.: Optimal and robust kernel algorithms for passive stochastic approximation. IEEE Trans. Inf. Theory **38**(5), 1577–1583 (1992)
102. Nemirovski, A., Juditsky, A., Lan, G., Shapiro, A.: Robust stochastic approximation approach to stochastic programming. SIAM J. Optim. **19**(4), 1574–1609 (2009)
103. Nesterov, Y.: A method of solving a convex programming problem with convergence rate o $(1/k^2)$ o (1/k2). Sov. Math. Dokl **27**, 372–376 (1983)
104. Niewiadomska-Szynkiewicz, E.: Application of evolutionary strategy to price management problem. In: Proceedings of VIII Conference on Evolution Algorithms and Global Optimization, KAEiOG (2005)
105. Nocedal, J., Wright, S.: Numerical Optimization. Springer Science & Business Media (2006)
106. Nocedal, J., Wright, S.J.: Sequential quadratic programming. Numerical Optimization, pp. 529–562 (2006)
107. Ogiela, M.R., Tadeusiewicz, R.: Syntactic reasoning and pattern recognition for analysis of coronary artery images. Artif. Intell. Med. **26**(1–2), 145–159 (2002)
108. Opara, K., Arabas, J.: Comparison of mutation strategies in differential evolution-a probabilistic perspective. Swarm Evol. Comput. **39**, 53–69 (2018)
109. Owens, D.H.: Iterative learning control (2015)
110. Owens, D.H.: Iterative Learning Control: An Optimization Paradigm. Springer, Berlin (2015)
111. Owens, D.H., Hätönen, J.: Iterative learning control–an optimization paradigm. Annu. Rev. Control **29**(1), 57–70 (2005)
112. Owens, D.H., Amann, N., Rogers, E., French, M.: Analysis of linear iterative learning control schemes-a 2d systems/repetitive processes approach. Multidimension. Syst. Signal Process. **11**(1), 125–177 (2000)
113. Owens, D.H., Hatonen, J.J., Daley, S.: Robust monotone gradient-based discrete-time iterative learning control. Int. J. Robust Nonlinear Control.: IFAC-Affil. J. **19**(6), 634–661 (2009)
114. Paszke, W., Aschemann, H., Rauh, A., Galkowski, K., Rogers, E.: Two-dimensional systems based iterative learning control for high-speed rack feeder systems. In: Proceedings of the 8th International Workshop on Multidimensional Systems (nDS), 2013, pp. 1–6. VDE (2013)
115. Pearson, J., Sridhar, R.: A discrete optimal control problem. IEEE Trans. Autom. Control **11**(2), 171–174 (1966)
116. Perry, L.A., Montgomery, D.C., Fowler, J.W.: Partition experimental designs for sequential processes: part i–first-order models. Qual. Reliab. Eng. Int. **17**(6), 429–438 (2001)

117. Perry, L.A., Montgomery, D.C., Fowler, J.W.: Partition experimental designs for sequential processes: part ii–second-order models. Qual. Reliab. Eng. Int. **18**(5), 373–382 (2002)
118. Perry, L.A., Montgomery, D.C., Fowler, J.W.: A partition experimental design for a sequential process with a large number of variables. Qual. Reliab. Eng. Int. **23**(5), 555–564 (2007)
119. Peto, J., Carpenter, J., Smith, G.D., Duffy, S., Houlston, R., Hunter, D.J., McPherson, K., Pearce, N., Romer, P., Sasieni, P., Turnbull, C.: Weekly COVID-19 testing with household quarantine and contact tracing is feasible and would probably end the epidemic. R. Soc. Open Sci. **7**(6), 200915 (2020)
120. Pimentel, M.A.F., Clifton, D.A., Clifton, L., Tarassenko, L.: A review of novelty detection. Signal Process. **99**, 215–249 (2014)
121. Polyak, B.T., Juditsky, A.B.: Acceleration of stochastic approximation by averaging. SIAM J. Control Optim. **30**(4), 838–855 (1992)
122. Polyak, B.T., Tsybakov, A.B.: Optimal order of accuracy of search algorithms in stochastic optimization (in russian). Problemy Peredachi Informatsii **26**(2), 45–53 (1990)
123. Postnikov, E.B.: Estimation of COVID-19 dynamics "on a back-of-envelope": Does the simplest SIR model provide quantitative parameters and predictions? Chaos, Solitons and Fractals **135**, 109841 (2020)
124. Poznyak, A.: Advanced Mathematical Tools for Control Engineers: Volume 1: Deterministic Systems, vol. 1. Elsevier, Amsterdam (2010)
125. Prates, M., Avelar, P.H.C., Lemos, H., Lamb, L.C., Vardi, M.Y.: Learning to solve NP-complete problems: a graph neural network for decision TSP. In: Proceedings of the AAAI Conference on Artificial Intelligence, vol. 33, pp. 4731–4738 (2019)
126. Price, K., Storn, R.M., Lampinen, J.A.: Differential Evolution: A Practical Approach to Global Optimization. Springer Science & Business Media (2006)
127. Priore, P., Ponte, B., Puente, J., Gómez, A.: Learning-based scheduling of flexible manufacturing systems using ensemble methods. Comput. Ind. Eng. **126**(September), 282–291 (2018)
128. Qu, G., Wierman, A.: Finite-time analysis of asynchronous stochastic approximation and $q$-learning. Proc. Mach. Learn. Res., TBD, 1–21 (2020)
129. Rafajłowicz, E.: Robustness of raw images classifiers against the class imbalance–a case study. In: IFIP International Conference on Computer Information Systems and Industrial Management, pp. 154–165. Springer (2018)
130. Rafajłowicz, E.: Classifying image sequences with the markov chain structure and matrix normal distributions. In: International Conference on Artificial Intelligence and Soft Computing, pp. 595–607. Springer (2019)
131. Rafajłowicz, E., Styczeń, K., Rafajłowicz, W.: A modified filter sqp method as a tool for optimal control of nonlinear systems with spatio-temporal dynamics. Int. J. Appl. Math. Comput. Sci. **22**(2), 313–326 (2012)
132. Rafajłowicz, E., Wnuk, M., Rafajłowicz, W.: Local detection of defects from image sequences. Int. J. Appl. Math. Comput. Sci. **18**(4) (2008)
133. Rafajłowicz, W.: Method of handling constraints in differential evolution using fletcher's filter. In: Rutkowski, L., Korytkowski, M., Scherer, R., Tadeusiewicz, R., Zadeh, L.A., Zurada, J.M. (eds.) Artificial Intelligence and Soft Computing, pp. 46–55. Springer, Berlin (2013)
134. Rafajłowicz, W.: Method of handling constraints in differential evolution using fletcher's filter. In: International Conference on Artificial Intelligence and Soft Computing, pp. 46–55. Springer (2013)
135. Rafajłowicz, W.: Numerical optimal control of integral-algebraic equations using differential evolution with fletcher's filter. In: Rutkowski, L., Korytkowski, M., Scherer, R., Tadeusiewicz, R., Zadeh, L.A., Zurada, J.M. (eds.) Artificial Intelligence and Soft Computing, pp. 406–415. Springer International Publishing, Cham (2014)
136. Rafajłowicz, W.: A hybrid differential evolution-gradient optimization method. In: Rutkowski, L., Korytkowski, M., Scherer, R., Tadeusiewicz, R., Zadeh, L.A., Zurada, J.M. (eds.) Artificial Intelligence and Soft Computing, pp. 379–388. Springer International Publishing, Cham (2015)

137. Rafajłowicz, W.: Learning novelty detection outside a class of random curves with application to covid-19 growth. J. Artif. Intell. Soft Comput. Res. **11**(3), 195–215 (2021)
138. Rafajłowicz, W., Jurewicz, P., Reiner, J., Rafajłowicz, E.: Iterative learning of optimal control for nonlinear processes with applications to laser additive manufacturing. IEEE Trans. Control Syst. Technol. **27**(6), 2647–2654 (2018)
139. Rafajłowicz, W., Rafajłowicz, E.: A rule-based method of spike detection and suppression and its application in a control system for additive manufacturing. Appl. Stoch. Model. Bus. Ind. **34**(5), 645–658 (2018)
140. Rafajłowicz, E., Rafajłowicz, W.: Iterative learning in optimal control of linear dynamic processes. Int. J. Control **91**(7), 1522–1540 (2018)
141. Ramaswamy, A., Bhatnagar, S., Quevedo, D.E.: Asynchronous stochastic approximations with asymptotically biased errors and deep multi-agent learning. IEEE Trans. Autom. Control, pp. 1–1 (2020)
142. Ramírez, A., Romero, J.R., Ventura, S.: A survey of many-objective optimisation in search-based software engineering. J. Syst. Softw. **149**, 382–395 (2019)
143. Reddi, S.J., Kale, S., Kumar, S.: On the convergence of adam and beyond (2019). arXiv:1904.09237
144. Reinhart, J.: Implementation of the response surface method (rsm) for stochastic structural optimization problems. Stochastic Programming Methods and Technical Applications, pp. 394–409. Springer, Berlin (1998)
145. Robbins, H., Monro, S.: A stochastic approximation method. Ann. Math. Stat. **22**, 400–407 (1951)
146. Robbins, H., Monro, S.: A stochastic approximation method. Ann. Math. Stat., pp. 400–407 (1951)
147. Rogers, E., Galkowski, K., Owens, D.H.: Control Systems Theory and Applications for Linear Repetitive Processes, vol. 349. Springer Science & Business Media (2007)
148. Rogers, E., Galkowski, K., Owens, D.H.: Feedback and optimal control. Control Systems Theory and Applications for Linear Repetitive Processes, pp. 235–304. Springer, Berlin (2007)
149. Rutkowska, D.: Neuro-fuzzy Architectures And Hybrid Learning, vol. 85. Physica (2012)
150. Rutkowski, L.: Flexible Neuro-fuzzy Systems Structure, Learning and Performance. Kluwer Academic Publishers, Dordrecht (2004)
151. Rutkowski, L.: New Soft Computing Techniques for System Modeling, Pattern Classification and Image Processing. Springer, Berlin (2004)
152. Nocedal, J., Wright, S.: Numerical Optimization. Springer, Berlin (2006)
153. Sachs, E., Guo, R-S., Ha, S., Hu, A.: On-line process optimization and control using the sequential design of experiments. In: 1990 Symposium on VLSI Technology, Digest of Technical Papers, pp. 99–100. IEEE (1990)
154. Sathya, N., Muthukumaravel, A.: A review of the optimization algorithms on traveling salesman problem. Indian J. Sci. Technol. **8**(1) (2015)
155. Scheinker, A., Krstić, M.: Model-free stabilization by extremum seeking. Number 9783319507897 (2017)
156. Schindele, D., Aschemann, H.: Norm-optimal iterative learning control for a pneumatic parallel robot. In: Gattringer, H., Gerstmayr, J. (eds.) Multibody System Dynamics, Robotics and Control, pp. 113–128. Springer, Vienna (2013)
157. Schittkowski, K.: More Test Examples for Nonlinear Programming Codes, vol. 282. Springer Science & Business Media (2012)
158. Schwabe, R., Walk, H.: On a stochastic approximation procedure based on averaging. Metrika **44**(1), 165–180 (1996)
159. Shen, X.N., Minku, L.L., Marturi, N., Guo, Y.N., Han, Y.: A Q-learning-based memetic algorithm for multi-objective dynamic software project scheduling. Inform. Sci. **428**, 1–29 (2018)
160. Shiue, Y.R., Lee, K.C., Su, C.T.: A reinforcement learning approach to dynamic scheduling in a product-mix flexibility environment. IEEE Access **8**, 106542–106553 (2020)

161. Shor, N.Z.: Nondifferentiable Optimization and Polynomial Problems, vol. 24. Springer Science & Business Media (2013)
162. Siemiński, A.: Verifying usefulness of ant colony community for solving dynamic tsp. In: Nguyen, N.T., Gaol, F.L., Hong, T.-P., Trawiński, B. (eds.) Intelligent Information and Database Systems, pp. 242–253. Springer International Publishing, Cham (2019)
163. Siemiński, A., Kopel, M.: Solving dynamic tsp by parallel and adaptive ant colony communities. J. Intell. Fuzzy Syst. (Preprint), 1–12 (2019)
164. Skubalska-Rafajłowicz, E.: Exploring the solution space of the euclidean traveling salesman problem using a kohonen som neural network. In: International Conference on Artificial Intelligence and Soft Computing, pp. 165–174. Springer (2017)
165. Skubalska-Rafajłowicz, E.: Random projection rbf nets for multidimensional density estimation. Int. J. Appl. Math. Comput. Sci. **18**(4), 455–464 (2008)
166. Spall, J.C.: A stochastic approximation algorithm for large-dimensional systems in the kiefer-wolfowitz setting. In: Proceedings of the 27th IEEE Conference on Decision and Control, pp. 1544–1548. IEEE (1988)
167. Spall, J.C.: A one-measurement form of simultaneous perturbation stochastic approximation. Automatica **33**(1), 109–112 (1997)
168. Spall, J.C. et al.: Multivariate stochastic approximation using a simultaneous perturbation gradient approximation. IEEE Trans. Autom. Control **37**(3), 332–341 (1992)
169. Sulikowski, B., Galkowski, K., Rogers, E., Owens, D.H.: Lmi based output feedback control of discrete linear repetitive processes. In: Proceedings of the 2004 American Control Conference, vol. 3, pp. 1998–2003. IEEE (2004)
170. Sun, H., Meinlschmidt, T., Aschemann, H.: Comparison of two nonlinear model predictive control strategies with observer-based disturbance compensation for a hydrostatic transmission. In: 2014 19th International Conference on Methods and Models in Automation and Robotics (MMAR), pp. 526–531. IEEE (2014)
171. Sutton, R.S., Barto, A.G.: Reinforcement Learning: An Introduction. MIT Press, Cambridge (2018)
172. Tanabe, R., Ishibuchi, H.: A review of evolutionary multimodal multiobjective optimization. IEEE Trans. Evol. Comput. **24**(1), 193–200 (2020)
173. Tátrai, D., Várallyay, Z.: COVID-19 epidemic outcome predictions based on logistic fitting and estimation of its reliability, pp. 1–15 (2020). http://arxiv.org/abs/2003.14160
174. Vidyasagar, M.: Learning and Generalisation: With Applications to Neural Networks. Springer Science & Business Media (2013)
175. Walk, H.: Foundations of stochastic approximation. Stochastic Approximation and Optimization of Random Systems, pp. 1–51. Springer, Berlin (1992)
176. Wang, B.C., Li, H.X., Li, J.P., Wang, Y.: Composite differential evolution for constrained evolutionary optimization. IEEE Trans. Syst., Man, Cybern.: Syst. **49**(7), 1482–1495 (2019)
177. Wang, D.J., Liu, F., Jin, Y.: A multi-objective evolutionary algorithm guided by directed search for dynamic scheduling. Comput. Oper. Res. **79**, 279–290 (2017)
178. Wang, I.-J., Chong, E.K.P., Kulkarni, S.R.: Equivalent necessary and sufficient conditions on noise sequences for stochastic approximation algorithms. Adv. Appl. Probab., pp. 784–801 (1996)
179. Wang, Y., Gao, F., Doyle III, F.J.: Survey on iterative learning control, repetitive control, and run-to-run control. J. Process. Control. **19**(10), 1589–1600 (2009)
180. Ward, R., Wu, X., Bottou, L.: Adagrad stepsizes: sharp convergence over nonconvex landscapes, from any initialization (2018). arXiv:1806.01811
181. Wasan, M.T.: Stochastic Approximation. Number 58. Cambridge University Press, Cambridge (2004)
182. Wiering, M.A., Otterlo, M.V.: Reinforcement learning. Adapt., Learn., Optim. **12**(3) (2012)
183. Wu, K., Darcet, D., Wang, Q., Sornette, D.: Generalized logistic growth modeling of the COVID-19 outbreak in 29 provinces in China and in the rest of the world, pp. 1–34, (2020). http://arxiv.org/abs/2003.05681

184. Xie, T., Yu, H., Wilamowski, B.M.: Neuro-fuzzy system. Intelligent Systems, pp. 20–1. CRC Press, Boca Raton (2018)
185. Bin Xin, L., Chen, J.C., Ishibuchi, H., Hirota, K., Liu, B.: Interactive multiobjective optimization: a review of the state-of-the-art. IEEE Access **6**, 41256–41279 (2018)
186. Xue, F., Sanderson, A.C., Graves, R.J.: Multi-objective differential evolution - algorithm, convergence analysis, and applications. In: 2005 IEEE Congress on Evolutionary Computation, IEEE CEC 2005. Proceedings, vol. 1, pp. 743–750 (2005)
187. Zou, F., Shen, L., Jie, Z., Zhang, W., Liu, W.: A sufficient condition for convergences of adam and rmsprop. In: Proceedings of the IEEE/CVF Conference on Computer Vision and Pattern Recognition, pp. 11127–11135 (2019)

# Index




W. Rafajłowicz, *Learning Decision Sequences For Repetitive Processes—Selected Algorithms*, Studies in Systems, Decision and Control 401,
https://doi.org/10.1007/978-3-030-88396-6

The manufacturer's authorised representative in the EU is Springer Nature Customer Service Centre GmbH, Europaplatz 3, 69115 Heidelberg, Germany. If you have any concerns regarding our products, please contact ProductSafety@springernature.com

Printed and bound by CPI Group (UK) Ltd, Croydon, CR0 4YY
30/09/2026
02225515-0003